Animal ecology in tropical Africa

Second edition

D. F. Owen

Longman
London and New York

Longman Group Limited London

Associated companies, branches and representatives throughout the world

Published in the United States of America by Longman Inc., New York

First edition published by Oliver & Boyd, 1966
Second edition published by Longman Group Ltd., 1976

Library of Congress Cataloging in Publication Data

Owen, Denis Frank.
Animal ecology in tropical Africa.

(Tropical ecology series)
Bibliography: p.
Includes index.
1. Zoology—Africa—Ecology. 2. Zoology—Tropics—Ecology. I. Title.
QL336.09 1976 591.5'0967 75–46586
ISBN 0–582–44363–6
ISBN 0–582–44362–8 pbk.

Set in 9 on 11 pt. Times New Roman by
Keyspools Limited, Golborne, Lancs
and printed in Great Britain by
Lowe & Brydone Printers Limited,
Thetford, Norfolk

Animal ecology in tropical Africa

TROPICAL ECOLOGY SERIES

Editors: D. W. Ewer and M. D. Gwynne

ALREADY PUBLISHED

Tropical forest and its environment
K. A. Longman and J. Jeník

Animal ecology in tropical Africa
D. F. Owen

FORTHCOMING TITLES

Ecology of tropical African mammals

Tropical soils

Acknowledgements

I am grateful to the following for permission to reproduce published text figures: the Editor of the *Journal of Ecology* (Fig. 1.10); the Editor, *Revue de Zoologie et de Botanique Africaines* (Fig. 2.8); the University of Chicago Press (Fig. 3.2); the Director of the National Museum, Nairobi (Fig. 3.3); the Editor of *Ibis* and Dr J. Green (Fig. 3.7); Uganda Shell Limited (Fig. 5.5); the Permanent Secretary of the Royal Swedish Academy of Science (Fig. 5.6); the Editor of *Nature* and Dr R. Hartland-Rowe (Fig. 5.7). Some of the material in Chapter 4 is adapted from my book *What is Ecology?* and I am grateful to Oxford University Press for permission to use it.

I thank the following for providing plates: Dr C. H. F. Rowell (Figs. 1.5, 3.10 and 3.13); Ministry of Information, Uganda (Figs. 1.6, 7.2 and 7.5); Professor O. Hedberg (Fig. 1.9); Professor T. L. Green (Figs. 1.7 and 4.3); East African Common Services (Fig. 2.1); Desert Locust Survey (Fig. 5.4); Dr Erasmus Harland (Fig. 7.1); British Information Services, Uganda (Fig. 7.4); John Blower (Fig. 7.6); De Beers Consolidated Mines Limited (Fig. 7.3); Grahame Dangerfield (Plates 1, 2 and 3), J. Hancock (Plate 14).

Cover illustration of the Fish Eagle (*Cuncuma vocifer*) drawn by Richard Bonson from a photograph by Arthur Christiansen supplied by Naturfoto, Denmark.

Contents

Preface to First Edition

My aim in this book is to provide an account of some of the ecological problems in tropical Africa. This is not a textbook, but rather an essay around a large and complex subject. As an essay I hope it will be useful to university students in tropical Africa and to more advanced students in schools. Visitors to Africa and some research organizations may also find something of value. I have not written extensively about 'big game', but rather have tried to select examples from the animal kingdom as a whole. Particular emphasis is placed upon peculiarly tropical problems, or on problems that can be best studied in the tropics.

Most of the book is about the work of others, but I have used my own work, chiefly in Uganda, where this seemed appropriate. The reader will find that birds, butterflies, and snails receive disproportionate attention. I make no excuse for this; these are the groups of animals that I happen to know best. The book was written at Makerere University College, Uganda, and I had available only those books, papers, and journals in the libraries of the College. As a result some older works were unavailable, as were most publications in languages other than English.

Some important aspects of ecology are hardly mentioned. The flow of energy between communities is receiving a great deal of attention from ecologists working in the temperate regions, but has not been investigated in tropical Africa. I have not stressed agricultural or medical problems, partly because these have already been or are being treated by others, but I have included a chapter on man. I have excluded mathematical treatment because this has repeatedly appeared in books concerned with general ecology, and as far as I know there are no major mathematical formulations that are peculiar to tropical Africa.

Where possible complex technical terms are avoided. A few are unavoidable and they are defined in the Glossary at the end of the book. Units of measurement are as given in the original work.

The text figures were drawn by Jennifer Owen. I am grateful to C. H. F. Rowell and Jennifer Owen for many valuable comments, and to June P. Thurston and L. C. Beadle for some useful suggestions. A grant from the University of East Africa covered the cost of preparing text figures and typing the text.

Kampala, 1965 D. F. Owen

Preface to Second Edition

The book has been completely revised and expanded, particularly by the addition of a chapter on communities and ecosystems (Ch. 4) but, as in the first edition, it is still an essay around a large and complex subject. There is now a vast literature on animal ecology in tropical Africa and the references I cite are no more than examples of what is available. I apologise for omissions; most of them are deliberate, but I have taken the opportunity of discussing some of the excellent work in French, especially from the Ivory Coast. The last chapter, on human ecology, reflects my own viewpoint and I trust it will be read as such.

I have converted most units of measurement to the metric system. Once again I thank Jennifer Owen for a critical reading of the text and for drawing additional figures. I thank also D. W. Ewer and Robert Welham (of Longman) for useful discussions and for some interesting ideas, and Angus McCrae for suggesting a wide range of improvements.

Leicester, 1975 — D. F. Owen

List of Colour Plates

Chapter 1

The biological scene now and in the past

This book is about Africa between 15°N. and 15°S. of the Equator, but I have not been too rigid in this restriction and there are a few references to situations outside the area, especially to the south. The area includes some of the largest and deepest lakes in the world and mountains rising to nearly 6 000 m. There are great rivers and extensive permanent swamps. Tropical Africa is an ecologically varied area but two major components dominate: the forest and the savanna, and throughout the book these are compared and contrasted.

To the south and, especially, to the north of the area under consideration there are vast deserts, but these are excluded from detailed discussion as are coastal and marine environments; I am concerned essentially with the ecology of terrestrial and fresh-water animals.

The northern limit of my area is the dry and barren Sahara Desert, the southern edge of which has recently suffered a severe drought that has caused much human misery. The desert acts as a barrier between the fauna of tropical Africa and the essentially European fauna of North Africa. The southerly boundary is less well-defined, as elements of the tropical African fauna extend south to the Cape. The east and west are bounded by the ocean. Fig. 1.1 is a political map of Africa and Fig. 1.2 a map of the main vegetation zones.

Ecology as a scientific study is concerned with the complex relationships between plants and animals and their surroundings, how they interact with one another, and how their numbers are limited by the availability of resources. All animals – birds, mammals, fish, insects, and the rest – are directly or indirectly dependent on plants, and although I have restricted this book to animal ecology I shall repeatedly make reference to vegetation and to the factors that determine its abundance and variety.

Present climate and vegetation

The most important climatic difference between the tropics and the temperate regions is caused by the greater altitude of the sun at midday. On the Equator at midday the sun is never more than $23\frac{1}{2}$ degrees from the vertical. It reaches this low angle at the solstices in December and June, and is directly overhead at the equinoxes in September and March. This means that in those parts of tropical Africa where there are seasonal changes, events are often repeated twice a year. Seasonal changes in mean daily temperature are small, often no more than a few degrees at the Equator, and are far smaller than the fluctuation experienced during any one day. The most important single climatic factor in the tropics is the amount and regularity of rainfall. In general, rainfall is of greater ecological importance than temperature in a tropical environment. Much of tropical Africa has a heavy annual rainfall, often exceeding 100 cm a year, and in areas where the rain is evenly distributed throughout the year there are biological environments such as evergreen forest, which, at least superficially, show little seasonal change. The generally drier savanna country experiences heavy seasonal rain, and in even drier semi-desert or desert country rainfall may be irregular and scarce. Generally speaking, the closer to the Equator the more even is the

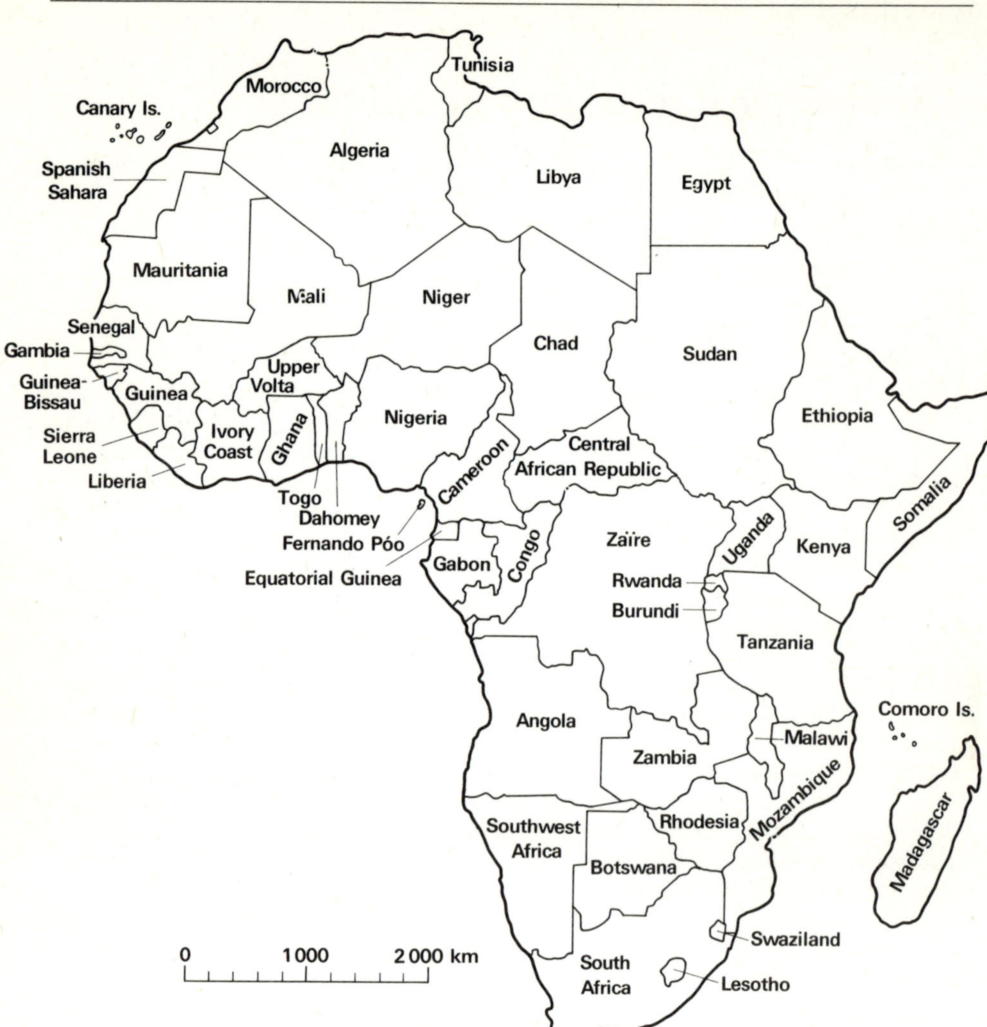

Fig. 1.1 Political map of Africa.

distribution of rainfall throughout the year, but there are many variations from place to place. Large lakes and mountains have a considerable influence on the rainfall pattern in the surrounding areas. And, of course, elevation above sea level has a profound effect on the climate and on the flora and fauna. The climate, vegetation, and animal life on high mountains in some ways parallel those of high latitude regions, but there is a major difference: at high latitudes the most important climatic fluctuations are seasonal, whereas high on tropical mountains the chief fluctuations, at least in temperature, are on a 24-hour basis.

In the vegetation map (Fig. 1.2) no account is taken of cultivation and destruction of the original vegetation by man. The forested areas include many different types of forest, such as deciduous, semi-deciduous, and evergreen. The humid, mostly evergreen, equatorial forest of west and central Africa reaches its eastern limit (apart from isolated patches) at the western section of the Rift Valley. Most of the forests in the east are high on mountains or around

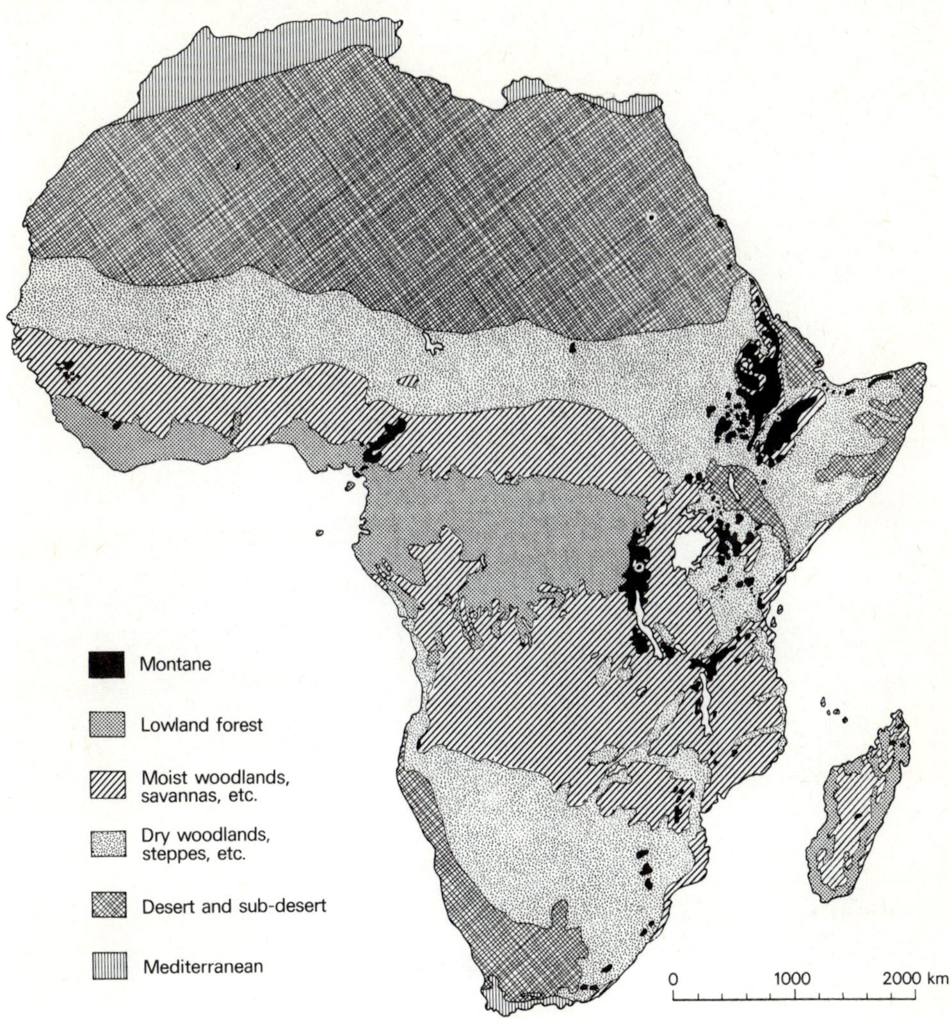

Fig. 1.2 Vegetation map of Africa.

permanent water. The central and western forest region is surrounded by moist woodland or wooded savanna, which, as shown in Fig. 1.2, gives way to dry woodland and steppe. All of this vast area, collectively called savanna, is dominated by grass. The savanna ranges from humid to arid and there is considerable variation in the vegetation as a result of these climatic differences. Grassy, treeless plains occur within the savanna and where it merges with the forest it gradually loses its identity (Fig. 1.3).

Many of the clumps of trees and shrubs that are so characteristic of savanna surround an old termite mound. All stages in the colonization of termite mounds by plants can be found, from a sparse covering of grasses and herbs to a dense tangle of woody shrubs and trees (Fig. 1.4). It is evident that termites are extremely important in determining the distribution of woody vegetation in the savanna. Since termites carry vast quantities of dead vegetation into the underground galleries of their mounds, they may create a small island of rich soil in the otherwise

Fig. 1.3 Where the forest meets the savanna there is a complex mixture of plant and animal species, many of them characteristic of neither community. The flowering bush in the foreground is *Lantana camara*, introduced from Central America and now well-established in tropical Africa.

Fig. 1.4 A large termite mound and its associated vegetation. The photograph was taken on the north shore of Lake Victoria in a region that could be described as 'man-modified forest edge.'

relatively barren savanna. As long as the termite mound is active, damp soil is continually brought to the surface for the construction of new galleries thus providing a favourable site for the germination of seeds, many of which may have been brought by the termites themselves or by birds and monkeys that like to sit on the tops of the mounds.

For more detailed vegetation maps than the one shown in Fig. 1.2, see Phillips (1959) and Keay (1959). Figure 1.5 shows a general view of a path inside mixed semi-deciduous and moist evergreen forest, while Fig. 1.6 is a view of savanna showing clumps of vegetation associated with termite mounds.

Throughout most of tropical Africa the influence of man on the vegetation can be readily seen (Fig. 1.7). Forests are being felled and cash crops such as coffee and sugar and staple crops such as plantains and cassava are replacing the original trees. Tropical forest can regenerate in 50 years or so, but once a forest has been felled there is a tendency for continued exploitation. The savanna is burnt over frequently in the dry season, and in many areas is heavily grazed by domestic animals. Cultivation extends well up the slopes of mountains, and soil erosion is of widespread occurrence in all these environments. Much of tropical Africa is now secondary bush consisting of a mixture of the original vegetation, introduced crops and their weeds, and elements of the vegetation of other environments. It is no longer easy to say with certainty whether a given area was once savanna or once forest.

The soil cover of Africa is varied. The soil develops primarily from the weathering of rocks, but also includes quantities of humus derived from decomposing vegetation. The rate of

Fig. 1.5 A path through the Mabira Forest, Uganda. Many of the trees are over 30 m high and are covered with lianes and epiphytes. This forest is largely evergreen but some trees are deciduous, losing their leaves in dry weather. Many trees have conspicuous buttresses. Forests like this teem with insects and birds, and monkeys of several species are conspicuous; other large mammals are rare or absent. (Photograph by C. H. F. Rowell.)

decomposition in warm and humid climates being high means that most of the nutrients are locked up in living plants. Removal of the plants not only removes the possibility of humus formation but also encourages erosion by wind and rain, leaching, and the formation of hard

Fig. 1.6 Savanna in western Uganda. The scattered clumps of trees and bushes mostly originate from termite mounds, which provide good conditions for seed germination and survival. There is a euphorbia tree on the right; these and acacias are the most common trees in this type of savanna. The bushes tend to be thorny and they are resistant to the fires that periodically destroy the grass. Savanna such as this can support large populations of ungulates and their predators. (Photograph by Ministry of Information, Uganda.)

Fig. 1.7 A view of the forest region of Sierra Leone. In the foreground are cultivated bananas and mango trees and there are a few oil palms in the right of the picture. The extent of previous cultivation can be seen about half-way up the slope by a line of demarkation between recent regeneration of the forest and forest that has not been cut for a long time, if at all. The tall palm contains many nests of the weaver bird, *Ploceus cucullatus*. (Photograph by T. L. Green.)

lateritic rock which is difficult if not impossible to cultivate. These processes help to explain why agricultural methods developed in temperate countries often fail when transferred to the tropics.

Past changes

During the past 60 million years tropical Africa has probably enjoyed greater stability of climate and geology than most other large areas of the world. The glaciations that occurred in the northern regions in the past million years undoubtedly had their effect on the fauna and flora of tropical Africa, but the effect was probably less marked than elsewhere. The few degrees drop in temperature that must have occurred during the northern glaciations would reduce evaporation and increase the biological effectiveness of rainfall. As a result the area of forest would have increased considerably. Many areas of isolated mountain forest in East Africa may have been joined together and also joined to the main west and central equatorial forest at various times during the past million years. Such repeated joining and fragmentation of the forest must have had important evolutionary and ecological consequences as populations were continually split and joined together again. Lake Chad may within the past 10 000 years have been four times the size of present-day Lake Victoria (Moreau, 1952). There is no doubt from the fossil record that the area at present occupied by the Sahara Desert had a rich fauna of mammals, probably like that of the present East African savanna. Analysis of pollen grains recovered from cores of mud taken from glacial lakes in the Ruwenzori Range suggests that in the past 15 000 years the vegetation has changed as much in the mountainous areas as it has in the temperate regions (Livingstone, 1967).

On the other hand some species and genera of animals now confined to the tropics once occurred in what are now cold temperate regions; there have in the past been fruit bats, hyaenas, cheetahs, rhinoceroses, crocodiles, and ostriches in Europe. The area now occupied by London was once sub-tropical forest. It appears that tropical Africa did not experience the marked fluctuations that have occurred in the composition of the fauna of the temperate regions during the past 60 million years. There has, then, been an enormous amount of time to build up stable and complex biological communities and this long-term stability has undoubtedly had its effect on the composition of the present fauna.

The present fauna

Any attempt to characterize the fauna of a large area necessarily depends on the work of competent taxonomists and collectors. The terrestrial vertebrates are better known than most other groups of animals, and for this reason the present fauna can best be discussed by reference to these animals. But it should be remembered that more than three-quarters of the species of terrestrial animals in tropical Africa are insects and that in all groups except the grasshoppers, mantids, dragonflies, and butterflies there are probably more undescribed than described species.

Tropical Africa has a greater array of mammals than any other area of the world. It is particularly rich in large ungulates, primates, and carnivores. The mammals of tropical Africa have closer affinities with those of tropical Asia than with those of other parts of the world. Old-World monkeys, apes, pangolins, bamboo rats (Rhizomyidae), elephants, and rhinoceroses are at present found in Africa and Asia and nowhere else. Endemic African mammals include the giraffe, hippopotamus, and aardvark families, three families of insectivores, and six families of rodents. The African savanna is dominated by large numbers of antelopes of many species, and with them are large carnivores, including the big cats and scavengers like hyaenas. The lowland forest contains many monkeys and the chimpanzee; the gorilla occurs in the highland and lowland forests of west and central Africa. Both fruit-eating and insectivorous bats are abundant. There are no marsupials, and in this respect Africa differs from South America and Australia. There are no true moles, beavers, or bears, all of which occur in the north temperate region. In South America there are groups of mammals that ecologically replace tropical African groups; these include mammals especially adapted for eating termites yet unrelated to the African pangolins and aardvark, and monkeys of two families not represented in Africa.

The breeding birds of Africa are again more similar to those of tropical Asia than to those of other parts of the world. Thirty per cent of the genera in Africa south of the Sahara are also found in the Asian tropics, but only 2 per cent of the species are common to both areas (Moreau, 1952). Several families of birds are confined to Africa, including the guinea fowls (7 species), mousebirds (6 species), turacos (18 species), and wood-hoopoes (6 species). Four other families of birds, each with a single species, are confined to Africa. Compared with other tropical regions of the world, Africa is especially rich in bee-eaters, barbets, honeyguides, rollers, weavers, starlings, shrikes, larks, and sunbirds. It has nearly as many hornbills, broadbills, and babblers as tropical Asia but is relatively poor in fruit pigeons, parrots, kingfishers, trogons, and woodpeckers.

Some tropical African species are replaced by ecologically equivalent but taxonomically unrelated species in other areas of the world. Sunbirds are small, brightly-coloured passerine birds with long down-curved bills. They feed on the nectar of flowers and on insects associated with nectar. In the American tropics the nectar-feeders are the hummingbirds which are related to swifts, but in colour, superficial structure, behaviour, and ecology resemble the sunbirds.

The yellow-throated longclaw, *Macronyx croceus*, is a common bird in the savanna of tropical Africa. It is a ground-nesting species and has an unmistakable voice and flight. It is bright yellow on the underparts with a black band across the upper breast. In the prairies of North America meadowlarks, *Sturnella* spp., occupy the same kind of habitat as the longclaw. They are closely similar to the longclaw in colour and pattern as well as in behaviour and without doubt fill the niche occupied by the longclaw in Africa.

The vultures of the New World and the vultures of the Old World (including Africa) are closely similiar ecologically and in superficial structure. But the two groups differ in important anatomical details and are placed by taxonomists in different sub-orders. The orioles and flycatchers are also remarkably similar in Africa and the New World but are placed in separate families because of fundamental anatomical differences.

In summary, the birds and mammals of Africa have greatest affinities with those of tropical Asia. African species are frequently replaced by ecologically similar but unrelated species in other parts of the world, especially in North and South America.

Reptiles are essentially tropical animals. Most of the species that occur in temperate areas are closely related to tropical species. Turtles, land tortoises, chameleons and other lizards, snakes, and crocodiles occur in tropical Africa, but crocodiles in particular have been severely persecuted by hunters and are nowadays rarely seen except in national parks. There are relatively few agamid lizards (although the rainbow lizard, *Agama agama*, is widespread and abundant) but there are many more species of chameleons than anywhere else in the world.

There are many frogs and toads, but no newts and no salamanders in tropical Africa. The tree frogs of Africa belong to a different family from those of tropical America. The clawed toads, *Xenopus*, are confined to Africa, although fossils have been found in Brazil (Estes, 1975).

The fresh-water fish of Africa are extremely varied and include an endemic family the Mormyridae, one of the two families of fish that have electric organs used in spatial orientation, and an endemic order the Polypterini (Fig. 1.8). Lung-fish occur in the swamps and rivers and are represented by related genera in South America and Australia.

The past fauna

Fossils are rare or absent in the humid equatorial regions of the world, simply because conditions are unsuitable for fossilization. As a result little is known of the past history and distribution of forest animals. Fossils have been found in plenty in the drier savanna, particularly in East Africa,

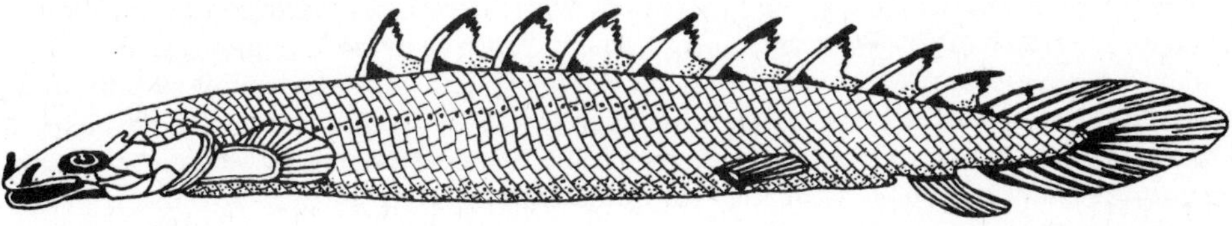

Fig. 1.8 *Polypterus bichir*, one of the several species of an endemic order of fish found in African rivers and swamps. These fish are relicts of an ancient group that first appeared about 400 million years ago.

and especially in association with fresh-water lakes and the dry beds of old lakes. They include aquatic and terrestrial molluscs, insects, fish, reptiles, birds, and mammals. There is fossil evidence of many apes in the savanna of East Africa, and it is possible that man originated in this area. Many fossil mammals are considerably larger than comparable modern species, although why this should be is not known. The distribution of certain species has changed in the past million years; several species of aquatic molluscs, fish, and crocodiles once occurred in parts of East Africa where today they are absent. Thus, in the Kazinga Channel of western Uganda there can be found great numbers of fossil crocodile teeth, but there are no living crocodiles in the area even though the habitat would appear suitable for them.

As already mentioned, the climate of the Sahara has in the past been much moister than at present. Throughout the Sahara there are traces of human habitation (rock paintings and engravings and the fossils of domestic animals), fossil elephants, rhinoceroses, hippopotamuses, and giraffes, indicating that within the past 10 000 years the area was much more suitable for animal life than at present.

It must be stressed, however, that the distribution of fossils reflects primarily the distribution of conditions that were suitable for fossilization. Thus, although fossil ape-men (or man-apes) have been found in parts of East Africa it cannot be stated with certainty that the major trends of evolution from ape to man were occurring only in East Africa. The equatorial forests, which have been stable biotic environments for so long, have left little record of the evolutionary changes that must have occurred within them.

Attempts have been made to relate the present distribution of animals in tropical Africa to past climatic changes. The birds are perhaps the best known for studies of this kind but the butterflies have also been used. On the whole the forest birds seem associated with particular vegetation types, and in some cases it is felt that this association can only be explained in terms of past events, in particular of past climatic changes (Moreau, 1963). The difficulty with this approach, which depends upon the existence of geographical variation at the subspecies and species level, is that virtually nothing is known about the rates of subspeciation and speciation in birds. It has been suggested that 5 000 years is the time taken for the evolution of a subspecies, but it is possible that this estimate is ten times too long – in any case no one can agree exactly what constitutes a subspecies.

The effect of isolation on the composition of the fauna

One important effect of the arrangement of rivers, mountains, and lakes and the pattern of vegetation (see Fig. 1.2) is that tropical Africa is split up into more or less isolated pieces. The fauna of the forest and the savanna have little in common and there is in many groups of animals, especially invertebrates, little interchange between these two major kinds of habitat. The effects of isolation on the composition of the fauna are best seen by consideration of two areas: high mountains and deep lakes, in which the discontinuity with surrounding environments is marked.

The several isolated mountain masses in tropical Africa, especially those in East Africa, present biological problems similiar to those encountered on oceanic islands. The mountains are islands isolated from each other by hot savanna or humid forest, and in most groups of animals there can be little interchange between adjacent mountains. Species of passerine birds such as white-eyes, *Zosterops*, and thrushes, *Turdus*, are represented by distinct subspecies on adjacent mountains in East Africa (Moreau, 1954), yet in some cases the distance between the mountains is no more than about 30 km. Subspeciation would not be possible unless there was complete or nearly complete isolation between these populations.

Mount Kilimanjaro is the highest mountain in Africa. It rises to nearly 6 000 m and lies just south of the Equator in northern Tanzania. The mountain is of volcanic origin, and at its base is about 80 km from east to west and 58 km from north to south. The hot thorn-bush savanna at its base is at an elevation of over 1 000 m; then, proceeding upwards, there is a cultivated zone extending to about 2 000 m followed by a zone of cloud forest to about 3 000 m. Above the forest there is moorland to about 5 000 m, followed by alpine desert with little vegetation; finally, there is little or no life, and the summit is under permanent snow and ice. It is one of the most spectacular mountains in the world. The moorland is occupied by giant species of heaths, groundsels, and related plants that are characteristic of the high mountains of East Africa (Fig. 1.9).

Fig. 1.9 Inside the crater of Mount Elgon at about 3 800 m above sea level. The plants in the foreground are *Helichrysum* spp., often called 'everlasting' flowers. The scattered 'trees' are the giant groundsel, *Senecio elgonensis*. (Photograph by O. Hedberg.)

The invertebrate fauna of upper Kilimanjaro has been described by Salt (1954). Some of the species found at high elevation are included in the temperate fauna of Europe and North

America, including three species of Collembola, three species of mites, and a fly. A few Kilimanjaro invertebrates are widespread in the lowlands of tropical Africa and a much greater number are found high on mountains elsewhere in East Africa, particularly on Ruwenzori, Elgon, and Kenya. Then there are species that are peculiar to Kilimanjaro; all but two of the 54 species collected by Salt and described as new to science have not yet been found elsewhere. In some cases this may be due to lack of collecting but the more conspicuous species, especially the carabid beetles, appear to be endemic. It is possible that the proportion of endemic species increases with altitude; of the 34 species from the alpine desert zone nearly two-thirds appear to be endemic.

A great many of the pterygote insects of upper Kilimanjaro have lost the ability to fly. Loss or reduction of wings is a feature of the insects of high mountains and oceanic islands; supposedly the increased windiness results in selection for wing reduction but the problem has not been properly studied. On Kilimanjaro there are flightless moths, beetles, earwigs, tipulid flies, and grasshoppers. A flightless tipulid, *Tipula subaptera*, is shown in Fig. 1.10. It is a remarkable insect, quite unlike the familiar lowland 'daddy long legs'. One evolutionary and ecological effect of flightlessness among high altitude insects is an increase in endemicity, as dispersal to adjacent mountains is prevented.

In the alpine desert zone of Kilimanjaro the temperature falls below freezing on every night of the year. The length of day remains nearly constant throughout the year, and animals living at this elevation are therefore subject to a remarkable but constant climate, unparalleled in the temperate regions or even in the lowland tropics; there is, in effect, winter every night and summer every day. No invertebrate could be nocturnal because it is always too cold at night. During the day the surface temperature is high and activity of invertebrates tends to be restricted to a few hours in the morning and in the evening. Hibernation and diapause are impossible as all stages of the life cycle are exposed to the complete range of climate every 24 hours. Hence an invertebrate living at this elevation must be adapted in physiology and in behaviour to this remarkable climate. Most species hide under stones during the extremes of cold and heat. The fact that the invertebrate fauna is relatively poor in species on high tropical mountains is perhaps explained by the peculiar climate of these regions; a scarcity of plants available for food may also contribute.

The fauna of the Rift Valley lakes is better known than the fauna of the high mountains. This is partly because the lakes are potentially of economic importance while the mountains are not, at least not biologically, and partly because there are many more fresh-water ecologists than there are mountain ecologists. The two biggest Rift lakes, Malawi (formerly Nyasa) and Tanganyika, resemble Lake Baikal in eastern Asia in their physical structure and in the great numbers of endemic species of animals in each. Lake Tanganyika, the larger of the two, may be considered from the point of view of the effect of isolation. The lake is in the southernmost and deepest section of the Western Rift and lies at an elevation of about 800 m. It is over 36 000 km^2 in area and about 1 400 m deep at its deepest known point. In places steep cliffs plunge directly into the lake. The lake had its origins about 10 million years ago and, so far as the fauna is concerned, has been isolated for 1 to 6 million years, depending on the kind of evidence used in making an estimate of this kind. Of a total of about 160 species, 136 endemic species of fish have been described from Lake Tanganyika, 42 of the genera being endemic. About 150 species of aquatic mollusc and 20 species of Crustacea are endemic. Endemic oligochaetes, flatworms, sponges, and Protozoa have also been described, but no estimate of the number of endemic species can at present be made because these groups have not yet been studied in detail. Many of the gastropod molluscs have thick ornamented shells reminiscent of marine species, and many of the fish look superficially like the species one might expect on a rocky sea-coast. But the fauna is probably not of marine origin and the similarities are the result of convergent evolution. Lake Tanganyika is a large enough expanse of water for the wind to whip up high waves; where these break on the shore, which in places is very rocky with precipitous cliffs, there is considerable resemblance to

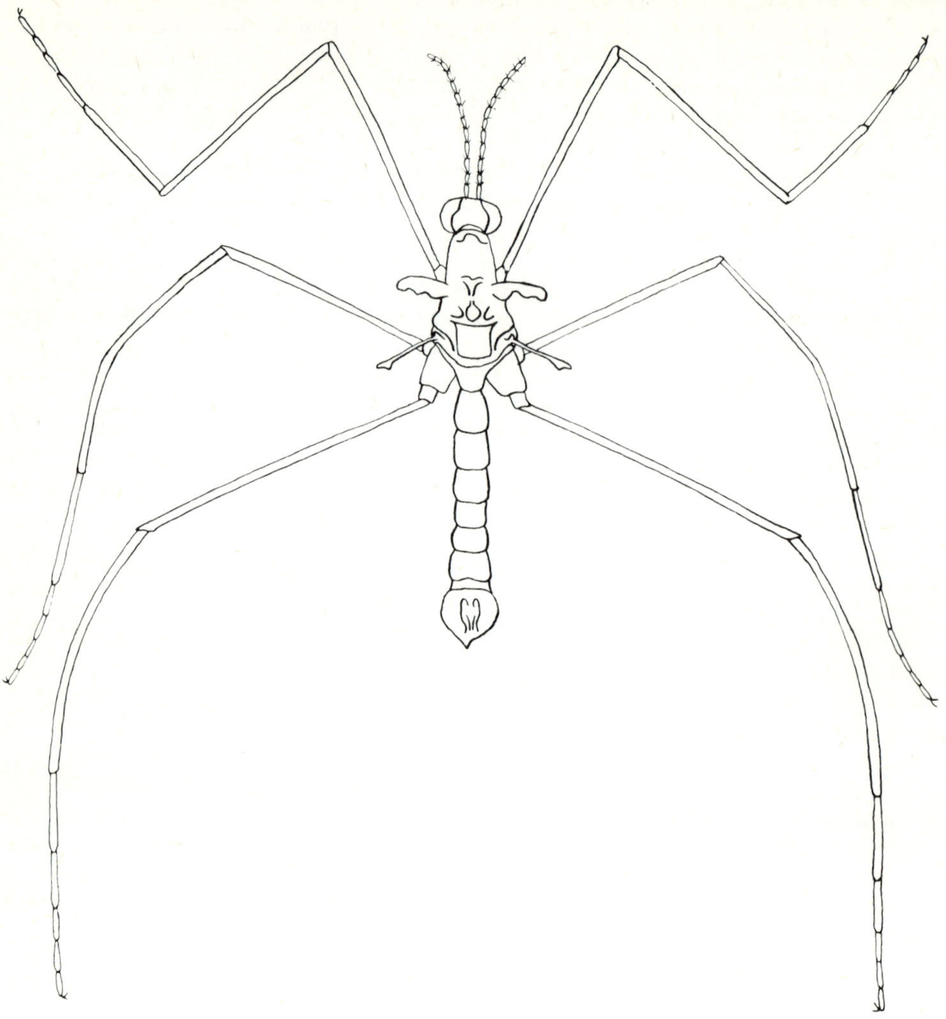

Fig. 1.10 *Tipula subaptera*, a flightless tipulid fly from the alpine desert of Mount Kilimanjaro. Most lowland tipulids have well-developed wings and can fly. Flightlessness is a feature of the pterygote insects of high mountains. (Redrawn from Salt, 1954.)

the sea. A more detailed account of the peculiar fauna of Lake Tanganyika is given in Brooks (1950).

Mount Kilimanjaro and Lake Tanganyika are examples of places with striking physical, climatic, and biological features that have resulted in their characteristic fauna. The number of endemic species, especially in the lake, is high and provides exceptional opportunities for investigations of species-composition, the effects of isolation, and the origin of species.

Evolutionary considerations

Taking the terrestrial and fresh-water animals as a whole, most of the species that occur in tropical Africa occur nowhere else in the world, except that many of them infiltrate the southern and more temperate part of the continent. It therefore seems likely that most of these species evolved in Africa south of the Sahara. The evolution of species is possible if populations become split and isolated from one another and if there is adaptation to local environments through natural selection. Tropical Africa undoubtedly provides many opportunities for speciation through isolation; the enormous numbers of endemic species in the East African lakes testify to this.

Evidence for speciation through isolation is necessarily circumstantial and must be deduced from the contemporary distribution of species. An excellent example is provided by the *Papilio demodocus* group of swallowtail butterflies. Throughout tropical Africa there is one common species of yellow and black, tailless swallowtail that is characteristic of open habitats; this is *Papilio demodocus*. There is a similar species, *Papilio demoleus*, considered by some to be the same species (they will breed together in captivity) in the Asian tropics. On the island of Madagascar four species of the *Papilio demodocus* group occur: *P. erithonioides*, *P. morondavana*, *P. grosesmithi*, and *P. demodocus* itself, the latter probably introduced by man in recent times. All four species are shown in Fig. 1.11. They are, as shown, similar in overall appearance, all being yellow with black markings, but in each species the markings differ in details. The differences between them are most easily seen by comparison of the shapes and positions of the outer ring of yellow spots in the hindwing and of the extent of development of tails.

The existence on Madagascar of four similar swallowtails in the *Papilio demodocus* group suggests successive invasions of Madagascar by butterflies from the mainland of Africa and subsequent evolution in isolation. Provided there is plenty of time between successive invasions the possibilities for speciation are considerable. It is not possible to place a time scale on the invasions nor is it possible to estimate the sequence of arrival, except that it seems certain that *Papilio demodocus* was the last one to become established. *Papilio demodocus* has a continuous distribution in Africa south of the Sahara and because populations are not isolated there has been no opportunity for speciation.

Another factor that may favour speciation in the tropics is the number of breeding generations possible in a given period of time. As an example, most European butterflies have one, two, sometimes three generations in a year, whereas tropical species may have twice as many; in West Africa and in the warmer parts of East Africa the monarch butterfly, *Danaus chrysippus*, may have as many as 12 generations in a year. But little is known of evolutionary rates in the tropics. In the temperate regions it is known that one allele can be replaced by another through natural selection in 50 to 100 generations, which may mean 50 to 100 years in an insect. Because of the increased number of generations it would seem that evolution could potentially be more rapid in the tropics than in the temperate regions. But evolution depends on environmental changes as well as genetic potentialities; environmental changes may on the whole be slower in the tropics than in the temperate zone. Evidently speciation could occur more rapidly in the tropics, but whether it does is a matter of conjecture.

The biological scene

What, then, does the animal ecologist find in tropical Africa? He sees vast forests full of animals, some brightly coloured, some cryptic. He sees the savanna, at times very dry and at times extremely wet. He is impressed by the importance of rain in the lives of most species, and by changes in temperature within each 24-hour period that are greater than seasonal changes. But most of all he is overwhelmed by the number and variety of species of animals, the subject of the next chapter.

Fig. 1.11 Four closely similar species of *Papilio* presumed to have evolved as the result of successive invasions of Madagascar from the African mainland. (1) *P. demodocus*, (2) *P. erithonioides*, (3) *P. morondavana*, and (4) *P. grosesmithi*. (From Owen, 1971.)

Chapter 2

The number, abundance, and diversity of species

A conspicuous feature of the tropical regions of the world is the large number of species of plants and animals. Many fewer species occur in the temperate regions. Humid tropical lowland forest is probably the richest habitat in the world as regards the number of species of plants and animals. Even a casual observer would note that a tropical forest contains far more species than a northern oak or spruce forest.

In 1878 Alfred Russel Wallace, best known for his association with Darwin in the formulation of the theory of evolution by natural selection, and who had vast experience in the tropical regions of the world, wrote:

> Animal life is, on the whole, far more abundant and varied within the tropics than in any other part of the globe, and a great number of peculiar forms are found there which never extend into the temperate regions. Endless eccentricities of form and extreme richness of colour are its most prominent features, and these are manifested in the highest degree in those equatorial lands where the vegetation acquires its greatest beauty and fullest development.

Wallace was particularly struck by the diversity of trees, and noted:

> If the traveller notices a particular species and wishes to find more like it, he may often turn his eyes in vain in every direction. Trees of varied forms, dimensions and colours are around him, but he rarely sees any one of them repeated. Time after time he goes towards a tree which looks like the one he seeks, but a closer examination proves it to be distinct. He may at length, perhaps, meet with a second specimen half a mile off, or may fail altogether, till on another occasion he stumbles on one by accident.

In these paragraphs Wallace appreciates three important features of plant and animal life in the tropics: the great number of species, the relative scarcity of many of them, and the presence of unusual species not found in other parts of the world.

In a northern European deciduous woodland there may be 20 species of tree and woody shrub to the hectare, while in eastern North America there may be two or three times this number. But in a hectare of lowland evergreen forest in Africa there may be as many as 300 species, and even more in an equal area in forests in South America and tropical Asia. Not only are there more species of trees in tropical forests but there is a much wider range of kinds of species; that is to say, there is increased diversity.

It is worthwhile at this point to clarify what is meant by the word 'diversity' in this context. To most ecologists it means the number of different species and when used in this way it is more appropriate to speak of 'species diversity'. But the word also means to stand apart in quality and in this sense need not necessarily refer to the number of species. Diverse species are those that differ in fundamental attributes from other species. Thus there may be in an area a great number of similar species but this does not represent a great diversity of species. There are, for example, many species of weaver birds in Africa, but they are rather similar in basic structure and hence there is not much diversity. On the other hand many of the African birds of prey are very different from each other in form and structure – in this group therefore there is considerable diversity. In this chapter I discuss diversity from both points of view.

The vegetation of an area has a profound effect on the number and abundance of animal species. In areas where there are many species of trees and plants there are usually many species of animals, and where there are few species of plants and trees there are generally fewer species of animals. This is because each species of animal is dependent, directly or indirectly, upon a relatively narrow range of plant species. Some insects feed on only one species of plant; many are confined to a single genus or family; only a few are able to exploit a wide range of plants. Predators, likewise, tend to be restricted to a relatively narrow range of prey species, but there are exceptions to this rule.

It is useful at this point to compare the number of species in some selected groups of animals in tropical Africa with comparable groups in temperate regions. It may then be possible to account for the greater number of tropical species in terms of the availability of resources, particularly of food. Five groups – mammals, birds, snakes, lycaenid butterflies, and hawk-moths – will be discussed in this way.

1. The mammals of Tanzania and Britain

There is a reasonably complete check-list of the land mammals of Tanzania in Swynnerton and Hayman (1950–51). With about 289 species this area possibly has more than any other area of comparable size and undoubtedly there are many smaller species, particularly bats and rodents, awaiting discovery. By comparison there are about 47 species of native land mammals in Britain, although several other species have been exterminated in the last few hundred years and a few have been accidentally or deliberately introduced from elsewhere.

There are 89 species of small herbivores in Tanzania that feed chiefly on leaves, stems, and seeds (especially grass seeds). They include 57 species of rats and mice and 10 of squirrels. Some of them undoubtedly take other foods but all are essentially plant-feeders. Forty-seven large herbivores occur, including 37 species of antelope and their relatives. With the exception of four species of pig (which take some animal food) all are directly dependent on plants for food. There are 91 insectivores, of which 56 are bats and 19 are shrews. Some, such as the pangolins which feed upon termites and ants, have a very restricted diet but many, like the bats that take nocturnal insects, are opportunists and exploit a wide variety of food items. There are 41 essentially flesh-eating carnivores, but some (hyaenas, jackals) are in part scavengers from bigger carnivores, and some (mongooses and their relatives) eat insects and fruit as well as flesh, while others eat fish, molluscs, and crabs. Eleven species of fruit bats eat fruit (including cultivated varieties) and pollen. Some of the 10 species of monkeys are fruit-eaters, others eat leaves – some, such as the baboon, are partly flesh-eaters. The chimpanzee is essentially a fruit-eater but occasionally takes insects, especially termites.

It must be pointed out, however, that apart from grass-feeders, many mammals, particularly monkeys and the smaller carnivores, are opportunist in regard to feeding, and although they can be broadly classified allowance must be made for variation.

Aside from the smaller number of species, the mammal fauna of Britain differs from that of Tanzania in: the absence of antelopes and the presence of three species of deer; the absence of essentially tropical mammals, such as monkeys, fruit bats, pangolins, aardvark, mole-rats, hyraxes; and the absence of the large herbivores and carnivores, but it should be noted that both these groups have in the past million years been common and widespread in the north temperate region, including Britain.

The abundance and diversity of insects and fruit and the year-round availability of plant food (especially grass) probably accounts for the large number and variety of species of mammals in Tanzania. Past history is also important: the environment of Tanzania has been relatively stable for much longer than that of northern Europe.

2. The breeding birds of Kenya and of France and Switzerland

About 802 species of land birds breed in Kenya, whereas about 241 species breed in the combined area of France and Switzerland. These species are distributed among 73 families in Kenya and 50 in France and Switzerland. Thirty families found in Kenya do not occur in France and Switzerland, including guinea fowls, turacos, wood-hoopoes, hornbills, barbets, honey-guides, cuckoo shrikes, bulbuls, and sunbirds, to mention only the larger ones. Seven families that occur in France and Switzerland do not occur in Kenya, but each contains relatively few species: these are grouse, oystercatchers, nuthatches, dippers, treecreepers, wrens, and accentors. Some of the families that occur in both areas are poorly represented in France and Switzerland: the weavers, including sparrows, are one such family.

There are, then, about three and a half times as many species of birds in Kenya as in France and Switzerland. What can be said to characterize the bird fauna of Kenya – a large and variable tropical area – compared with that of France and Switzerland – a large and variable temperate area? In Table 2.1 are listed some of the families of birds that contain a total of 10 or more species in the two areas combined. These families are selected in an attempt to demonstrate the difference in the bird fauna of the two areas.

Table 2.1 *The number of species in some of the larger families of birds found in Kenya and in France and Switzerland.*

	Estimated number of species	
	Kenya	*France and Switzerland*
Herons	12	7
Ducks and geese	14	13
Birds of prey and vultures	40	17
Game birds	17	5
Rails and crakes	14	7
Pigeons and doves	15	5
Turacos	10	–
Cuckoos	14	2
Owls	12	8
Nightjars	11	2
Swifts	11	3
Kingfishers	10	1
Hornbills	12	–
Barbets	19	–
Woodpeckers	12	8
Swallows	13	4
Crows	4	9
Titmice	7	8
Bulbuls	23	–
Warblers	64	27
Flycatchers	34	3
Shrikes	32	4
Sunbirds	32	–

Birds that are mainly predators of vertebrates are better represented in Kenya. Thus herons, birds of prey, owls, and kingfishers are represented by two or three times as many species. Many of the kingfishers of Kenya are predators of terrestrial vertebrates and large insects, but there are also some fish-eating species; many of the birds of prey are scavengers and feed chiefly on dead mammals. Fruit- and seed-eaters, such as pigeons and turacos, are represented by five times as many species in Kenya. Predators of large insects, such as shrikes, have about eight times as many species – sunbirds, which are nectar-feeders and have 32 species in Kenya, do not occur in France and Switzerland. There are many more species of birds in Kenya that feed upon winged insects; these include swifts, nightjars, swallows, flycatchers, and many of the warblers, but caterpillar- and nut-feeders, such as titmice, are less well represented. Ducks and geese are equally represented in the two areas; many are herbivores and some are grazers. Crows are omnivores though in France and Switzerland the species are at least partly dependent upon earthworms and caterpillars; earthworms are not especially abundant in Kenya, which may explain the rather few species of crows. Woodpeckers are a puzzling group; a great many species might be expected in Kenya. They are not replaced by any other group of birds and their relative scarcity (in number of individuals as well as in species) is possibly because most of the trees have smooth bark and are therefore less likely to harbour large numbers of insects.

Kenya, then, differs from France and Switzerland in having many more predatory and insectivorous birds, more nectar- and fruit-eating species, and a paucity of species that eat earthworms.

But comparisons such as this involve many practical difficulties. Kenya and France and Switzerland have been considerably modified by man. The effect of this on the bird fauna is to make many habitats unsuitable for certain species while other species flourish. Another difficulty is that the grazing birds are in Kenya to some extent replaced by grazing mammals, and it may well be misleading to restrict the comparison to a single taxonomic group – perhaps, for example, woodpeckers in Kenya are replaced by large predatory insects. Lastly, many of the species that breed in France and Switzerland spend only half the year there. Many such species migrate and spend the northern winter in Africa, including Kenya. The effect of this annual influx of northern migrants on the species-composition of the resident bird fauna has hardly been explored, but it must be considerable.

3. The snakes of a cocoa farm in Ghana

A total of 91 species of snakes has been recorded from Ghana and 34 species have been found in a sample of 203 snakes collected on a small cocoa farm (Leston and Hughes, 1968). In Ghana, as everywhere else in Africa, there are many more species of snakes than in temperate regions; only three species occur in the whole of Britain. The snakes recorded on the cocoa farm are almost all forest species and there is little to suggest penetration of this habitat by savanna snakes, as has occurred elsewhere in man-modified environments in West Africa (Menzies, 1966). Most of the species were represented by rather few individuals; 9 species were recorded once only, and 27 were recorded less than 10 times each. The commonest species was recorded 27 times.

This pattern of species-abundance occurs again and again in samples of plants and animals: there are a few common species and many relatively rare ones; indeed the distribution of individuals per species approximates to a logarithmic series. Because of this it is possible to predict how many additional species of snakes may be expected if the sample size is increased (Williams, 1964). Thus if the sample from the cocoa farm were increased to 400 an additional 7 species could be expected, but to obtain 91 species (the number known from the whole of Ghana) the sample size would have to be increased to about 30 000. This of course would be virtually impossible, but if such a sample were obtained it would presumably add several species to the Ghana list as it is likely that some of the species known from Ghana (especially from very dry areas) would never occur on a cocoa farm.

4. The diversity of lycaenid butterflies in small areas

Butterflies of the family Lycaenidae (hairstreaks, coppers, blues, and their relatives) are small- or medium-sized insects that occur in almost all parts of the world. Many are brilliantly coloured; others are more sombre and live in the deep shade of tropical forests. About eight species might occur on a small patch of chalk grassland in southern England, and about six in some oakwoods. One, two, possibly three species might be expected in an average English garden.

Africa is extraordinarily rich in lycaenids; many are rare, and the life histories of most species remain undescribed. Two collectors stationed periodically on platforms at various levels on a 36-metre high steel tower (Fig. 2.1) erected in the Mpanga Forest, Uganda, obtained a total of 78 species in a period of 9 months (Jackson, 1961). This total, from a very small area of forest, is 10 more than the species of butterfly in the whole of Britain and emphasizes the remarkable variety of lycaenids in Africa. The majority of the species collected were members of the subfamily

Fig. 2.1 A 36-metre high steel tower erected in the Zika and Mpanga Forests in Uganda by the East African Virus Research Institute, Entebbe. There are platforms at regular intervals upon which insect traps have been placed and from which observations of insect activity are made. The tower has been used to study insect flight activity at various levels in and above the forest. (Photograph by East African Common Services.)

Lipteninae, which are rather small weak-flying butterflies mostly confined to forests. They fly near the canopy and the females lay their eggs on tree trunks. The larvae feed on lichens and are intimately associated with ants. The ants seem to protect them from predators and in exchange receive tiny droplets of a sugary secretion which the larvae produce from a special gland. In the Mpanga Forest, and elsewhere in Africa, the diversity of lycaenids may be partly because of the very large numbers of species of ants.

For nearly 4 years I collected butterflies in two gardens near Freetown, Sierra Leone. Altogether I found 62 species of Lycaenidae – 42 in one garden, 37 in the other, and 17 occurred in both gardens. The gardens are in the forest region of Sierra Leone but both have small open areas of grass, and although most of the lycaenids collected were forest species, the most abundant species were invaders from the savanna. In all, 217 species of butterflies were recorded from one garden and 198 from the other but additional species, most of them lycaenids, were being added (at the time I left Sierra Leone) at about one a week. I think that at least 300 species of butterflies could be recorded from each of these gardens.

5. Hawk-moths in gardens

In all parts of the world gardens are attractive places for insects that feed from the nectar of flowers; indeed gardens often provide a greater concentration of flowers than even the most exotic natural habitats. One group of insects, the hawk-moths (Sphingidae) are especially attracted to garden flowers. Between four and six species may be expected in an English garden, and about three times as many in a garden in the eastern United States.

Most hawk-moths fly in the early evening or at night and can be readily caught in specially designed light-traps fitted with a mercury vapour bulb. For 20 consecutive months I operated a light-trap in a garden in Sierra Leone. A total of 6 619 individual hawk-moths of 52 species was obtained, which may be compared with the 51 individuals of 4 species obtained in the same period of time with a similar trap in a suburban garden at Leicester, England. Many of the species obtained in the Sierra Leone garden were relatively infrequent; 19 species were recorded less than 10 times each, and 8 appeared once only. But a few were exceedingly common, including the oleander hawk-moth, *Deilephila nerii* (Fig. 2.2), which was recorded 1 828 times. One species new to science, recently described as *Phylloxiphia oweni* (Fig. 2.3), was collected in 1967 in a garden in Sierra Leone, demonstrating that even in a well-known group of insects like the hawk-moths there are discoveries to be made even in a garden.

The sample of hawk-moths, like the sample of snakes from the Ghana cocoa farm, shows what an enormous diversity of species may occur; many are rare, a few common, and there is always the possibility of adding further species to an existing list if sampling is continued.

The Ichneumonidae as an exception to the rule

The rule is that for most groups of organisms there is a gradient of decreasing species diversity from the equator to high latitudes. I have provided examples that support the rule from mammals, birds, snakes, butterflies, and hawk-moths. It would be possible to give further examples, from plants as well as from animals, but I shall instead discuss an apparent exception that has recently come to light.

The Ichneumonidae are small, wasp-like, parasitic members of the order Hymenoptera. Virtually all species feed as larvae inside the larvae of insects which undergo a complete metamorphosis, especially butterflies, moths, beetles, and sawflies. Ichneumonids are rarely noticed; indeed the majority of zoologists would have difficulty in recognizing them and yet there are probably more species than there are of vertebrates, including fish. In few groups of animals can there remain so many undescribed species (Fig. 2.4).

Ichneumonids are abundant wherever there is vegetation and where it is not too dry but few

Fig. 2.2 The oleander hawk-moth, *Deilephila nerii*, one of the commonest hawk-moths in Africa. It has a wingspan of 11 cm and is beautifully patterned with green, pink, and brown. The moth is a well-known migrant and occasionally occurs in northern Europe.

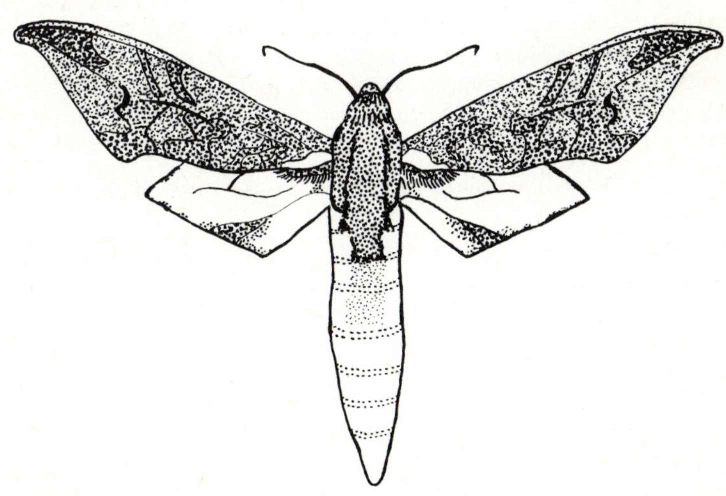

Fig. 2.3 *Phylloxiphia oweni*. It has a wingspan of 10 cm, the forewings are pale brown with a delicate pinkish shade and the hindwings are brick-red. Besides Sierra Leone the species is now known to occur in Liberia, the Ivory Coast, Nigeria, Cameroon, and Zaïre.

Fig. 2.4 Ichneumonidae. This specimen of *Owenus minor* is one of six species of *Owenus* I collected in a special trap in a garden in Uganda. Although I obtained a total of 129 specimens of *Owenus minor* I never saw it alive. The genus was described by Henry Townes, a world authority on this group of insects. The drawing shows only the left half of the specimen in a pinned and therefore unnatural position. The insect is about 8 mm long.

species seem to be common. Large and representative samples of them have been obtained from three gardens: at Kampala in Uganda, Freetown in Sierra Leone, and Leicester in England. Another sample, not from a garden, but from disturbed land along a stream, has been collected in Skåne in southern Sweden. The three gardens are not of course ecologically identical – Kampala and Freetown are on opposite sides of tropical Africa; the garden at Leicester is in the suburbs of a large city – nevertheless they share features common to all gardens: an abundance of flowers and a wide variety of plants and trees introduced from other parts of the world. The Skåne sample should perhaps be considered separately and is added only to provide further evidence of the pattern of species diversity discovered.

Results of an analysis of the four samples are given in Table 2.2. The differences in sample size

Table 2.2 *Two tropical and two temperate samples of Ichneumonidae compared. (From Owen and Owen, 1974.)*

	Sample size	*Species*	*Species taken once*	*% commonest species*
Kampala	2 268	293	116	10·1
Freetown	1 979	319	117	4·9
Leicester	2 495	326	122	3·2
Skåne	10 994	758	203	5·5

do not necessarily reflect differences in abundance; much depends on the habitat and the efficacy of the sampling method in different situations – although it should be noted that the same kind of trap was used in each locality – but I have the impression that, as a group, ichneumonids are rather more abundant in the temperate than in the tropical localities. All four localities produced an extraordinary number of species. At each roughly a third of the species was taken once only, which emphasizes particularly well the rarity of many species of ichneumonid as does the calculation of the percentage of each sample made up of the commonest species; indeed there were no really common species in these samples. It has been shown mathematically (Owen and Owen, 1974) that in these samples there is not the expected increase in species diversity in the tropical localities. The greatest diversity occurs in the Skåne sample while Kampala is the least diverse.

A large (but for the tropical species, unknown) proportion of the species of ichneumonid parasitize the caterpillars of Lepidoptera (butterflies and moths). There is no doubt about the greater diversity of the Lepidoptera in the tropics. I have already mentioned that there are probably about 300 species of butterflies and 52 species of hawk-moths in the Freetown garden, while in the Leicester garden there are 15 species of butterflies and only 4 of hawk-moths. These two groups comprise only a small fraction of the total species of Lepidoptera but probably reflect the situation in the remainder. Evidently there is no increase in diversity of ichneumonids with an increase in diversity of their hosts, contrary to what might have been expected. Why is this? The most likely explanation is that each species of ichneumonid tends to be niche-specific rather than host-specific; that is to say it seeks suitable hosts in certain well-defined places and is not adapted to a particular species of host. If this is so there may not be (for ichneumonids) a greater number of niches available in the tropics than in temperate regions, in contrast to the situation that probably prevails for most other feeding relationships in the two environments. Whatever the explanation it seems certain that the ichneumonids are an exception to a well-established trend.

The causes of high species diversity in the tropics

Plants and the animals that feed on them have evolved together. In general the greater the diversity of species of plants in an area the greater the diversity of plant-feeding animals and their predators and parasites. The high species diversity in most groups of animals that is so characteristic of the tropics is probably largely a consequence of high plant diversity but this itself may be at least partly associated with high animal diversity. Flowering plants that are pollinated by insects, birds, and bats could not have reached their present diversity without the evolution of pollinators: the high diversity of flowering plants in the tropics is largely the result of the high diversity of pollinators, and *vice versa*.

High animal diversity is obviously associated with high plant diversity but exactly what feature of the tropical environment causes this is a matter for conjecture. The longer favourable growing season in the tropics could allow more species to exploit available resources: the edible parts of plants are available for longer periods in the tropics than in temperate regions. Thus favourable climate could lead to increased diversity, or rigorous high latitude climates could reduce diversity. It is possible that in the tropics biotic interactions between species play the dominant role in regulating numbers, while in the temperate regions climatic factors are relatively more important. This could theoretically lead to smaller ecological niches in the tropics, which in plant-feeding insects might result in increased food specialization. Another view is that temperate regions are impoverished as a result of recent climatic catastrophes such as the Pleistocene glaciations. On this view the re-invasion of species that have become extinct at high latitudes is not yet complete. But this view is contradicted by the high diversity of ichneumonids in temperate regions which perhaps implies that plant diversity and year-round favourable climate are of prime importance – for if, as seems likely, ichneumonids are niche-specific they are probably less dependent on plant variety or on the prolonged availability of their plant-feeding hosts.

Unusually diverse species

Earlier in this chapter I suggested that diversity could also be understood in terms of species that differ in fundamental attributes and adaptations from other (related) species. A. R. Wallace repeatedly drew attention to the greater number of peculiar and unusual species of animals found in the tropics – especially in forests. Tropical representatives of many widely different animal groups differ markedly in structure from their closest relatives and some are highly specialized in their way of life. A few examples, chosen from different groups of animals, serve to illustrate the point.

In all the major tropical areas of the world there are mammals that feed mainly upon ants and termites. The pangolins, *Manis* spp., of Africa and tropical Asia are the sole representatives of a distinct mammalian order, the Pholidota. They feed on termites and to a lesser extent on ants and have many modifications in structure associated with this specialized diet. Figure 2.5a shows

Fig. 2.5 (a) A pangolin, *Manis* sp.; (b) An aardvark, *Orycteropus afer*. These mammals are termite-eaters, and have many structural modifications associated with this way of life. (Redrawn in part from Bere, 1962.)

the general appearance of an African pangolin. It has no teeth, the snout is long and there is a long tongue. The front feet have powerful claws for digging, and the eyes are small. Horny overlapping scales cover most of the body. Pangolins are nocturnal and some climb trees, but little is known of their numbers or ecology except that they are inconspicuous and eat termites and ants. They are very different from all other mammals.

Another ant- and termite-eater, the aardvark, *Orycteropus afer*, is the sole representative of the order Tubulidentata. It is confined to Africa, although fossil evidence suggests that aardvarks once had a much wider distribution in Europe and Asia. As shown in Fig. 2.5b it has a long snout, a small mouth, but, unlike pangolins, does not have scales. It has a long tongue and a few peg-like cheek teeth. Aardvarks feed at night, spending the day in a hole in or near a termite mound, but again little else is known of their ecology or numbers. They may be the only survivors of a much bigger order of mammals, which may even have ranged as far as North America.

A third species, the aardwolf, *Proteles cristatus*, is a member of the order Carnivora. It is essentially a specialized hyaena and feeds almost exclusively on termites. Compared to a hyaena it has a small head and shoulders and five rather than four toes on the front feet. It has 24 teeth (a hyaena has 34) and except for the canines, which are used for fighting, the teeth are rather small and peg-like. Unlike hyaenas, the aardwolf does not crush bones and this no doubt accounts for the reduced head musculature and weak teeth. The aardwolf is confined to arid areas in east and southern Africa and nothing is known from the fossil record of its past history. In the Serengeti National Park, Tanzania, the aardwolf feeds almost entirely on the termite, *Trinervitermes bettonianus* and related species (Kruuk and Sands, 1972) and in terms of diet must rank as one of the most selective and specialized members of the Carnivora.

Thus in Africa there are three types of mammals, each belonging to a separate order, that feed on termites and to some extent on ants and other insects. Each exhibits a number of convergent adaptations that must have evolved independently and which are associated with their diet.

The fossil record of birds is considerably less informative than that of mammals and little is known of the past distribution and relationships of some strikingly diverse African species. One of these, the whale-headed stork *Balaeniceps rex*, is placed in a family by itself, the Balaenicipitidae. The whale-headed stork is quite rare and is hardly ever seen unless a special search is made. It inhabits papyrus swamps in Uganda, the southern Sudan, and Zaïre, and looks like a large heron but, as shown in Fig. 2.6, its bill is enormous and shaped like a shoe. Like many

Fig. 2.6 An unusual-looking tropical African bird, the whale-headed stork, *Balaeniceps rex*. The presumably specialized function of its remarkable beak remains a mystery.

herons and some storks it feeds upon fish, including lung-fish and amphibians, but little is known of the function of the remarkable bill which is unparalleled in any other species of African bird.

The honeyguides, Indicatoridae, are a family of birds that occur in tropical Africa and in tropical Asia. They are forest species and frequently make themselves conspicuous by their behaviour and voice, thus attracting mammals, including man, to bees' nests. The mammal then destroys the nest for its honey and the honeyguide feeds on pieces of beeswax that are left scattered around. Apart from this remarkable behaviour, honeyguides are unusual in that they are among the few vertebrates that can digest quantities of wax.

The classification of mammals and birds is to a considerable extent based on anatomical adaptations associated with feeding. There are, however, a few species, especially among the birds, that are virtually impossible to classify with confidence. Among them are the two similiar species of rock-fowl, *Picathartes*, that inhabit West African forests. These birds look superficially like large starlings but have enormously developed legs and a head bare of feathers (Fig. 2.7). They build nests of mud in caves and on over-hanging cliff faces, often forming

Fig. 2.7 The rock fowl, *Picathartes gymnocephalus*, a bird of uncertain evolutionary relationships, confined to forests in West Africa.

colonies. Rock-fowl feed on large insects and in Sierra Leone I have several times noted that they remain near dense trails of driver ants and presumably catch insects disturbed by the ants as they move through the forest. The genus *Picathartes* has always been a puzzle to bird taxonomists. It has been placed in the crow, starling, and babbler families, and even in a family by itself. The last word about these strange birds has not been said although a recent study of the proteins present in their egg-white seems to suggest that they are babblers.

Tropical Africa has many diverse and unusual invertebrates. There is a caddis fly, *Limnoecetis tanganicae* (order Trichoptera), that as an adult skates over the surface of the water like a pond-skater (Hemiptera) or a whirlgig beetle (Coleoptera). It is confined to Lake Tanganyika and was

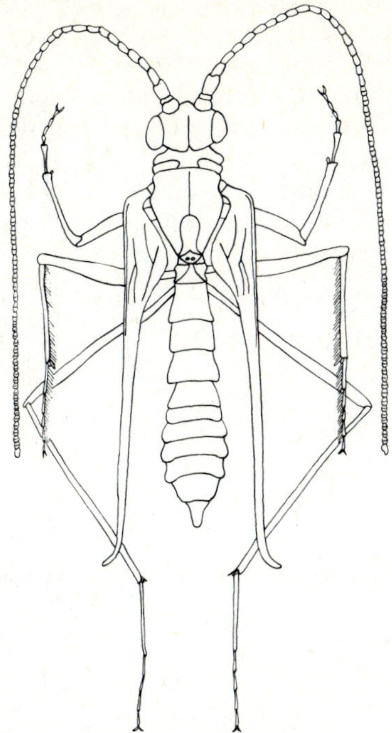

Fig. 2.8 The water-skating caddis, *Limnoecetis tanganicae*, of Lake Tanganyika. (From Marlier, 1955.)

first described as recently as 1955. The species is illustrated in Fig. 2.8, where it can be seen that it bears a strong resemblance to a bug rather than to a caddis. Then there is a reduviid bug, *Acanthaspis petax*, the nymphs of which decorate themselves with particles of soil and the corpses of their prey, especially ants. The bug lives on large termite mounds and can be recognized by what appears to be a cluster of dead ants moving quickly over the surface of these mounds. The larvae of a fruitfly, *Leucophenga* sp. (Drosophilidae), live in the masses of spittle formed by nymphs of the cercopid bug, *Ptyelus flavescens* (Odhiambo, 1958). In the East African lakes there is a wood-boring mayfly (Ephemeroptera) nymph; this species, *Povilla adusta*, has powerful mandibles associated with wood-boring, but it is a filter-feeder and does not eat wood.

Jellyfish are usually associated with the sea but some species occur in fresh water: in some lakes and rivers the medusae of *Limnocnida* spp. are occasionally found in dense swarms; the hydroids are attached to the stems of aquatic plants. Few people have reported *Limnocnida* from Africa and it is not certain whether there are one, two, or three species (Green, 1960).

In the oxygen-deficient swamps of tropical Africa there is an oligochaete worm *Alma emini*, which possesses an external longitudinal groove on the dorsal surface of the hinder end. This structure is a respiratory organ, and occurs in other worms that live in oxygen-deficient swamps, especially in South America (Beadle, 1933).

A jumping millipede seems an improbable animal, yet in Sierra Leone there is a species, *Diopsiulus regressus*, that jumps when disturbed (Evans and Blower, 1973). The usual defensive movement of millipedes is a rapid spiralling of the body to enclose the head and legs, and in *D. regressus* jumping is thought to be an alternative method of escaping a predator.

Then there is *Oculotrema hippopotami*, a parasite that lives on the surface of the eye of the hippopotamus. This parasite belongs to the order Monogenea and was first reported from the eye of a hippopotamus in the zoological gardens at Cairo. Parasitologists were extremely doubtful about the origin of the first specimen as the Monogenea are parasites of fish and amphibians. But it is now known that more than three-quarters of the hippopotamuses in Ruwenzori National Park, Uganda, are infected with *O. hippopotami*, the only species of the order that has been found on a mammal (Thurston and Laws, 1965).

Many additional examples of diverse and unusual tropical African animals could be cited. There are, perhaps, two kinds of diversity: unusual species, like pangolins, that are anatomically distinct and exploit a specific food source: and species, like the fruitfly *Leucophenga* sp., described above, that have unusual ways of life yet belong to widely distributed groups of animals.

Chapter 3

Populations

The word population is from the Latin *populus* which means people, and for a long time virtually all discussion of population concerned human numbers. Nowadays the word population is used to mean a number of organisms of the same species living and breeding together. It is inaccurate to speak of a population containing more than one species, although some ecologists persist in doing this.

The rates of reproduction of most organisms are extremely high and all populations are capable of rapid, exponential growth in size, yet despite this most populations remain stable. Ecologists are interested in the environmental factors that maintain stability, which, in effect, means that they are interested in the balance between birth and death rates. The size of a population is determined by the resources of the environment. Since these resources cannot always be specified it is useful to think in terms of what is called the carrying capacity of the environment. If, for instance, a population is growing exponentially it will, in normal circumstances, quickly reach the carrying capacity and an acute shortage of resources will begin to exert severe restraints on further growth. As the carrying capacity is approached the rate of population growth will slow down and soon the death rate will equal the rate of recruitment of new individuals into the population. Thereafter the population is regulated at a fixed level with only minor fluctuations in size compared to what is possible. The carrying capacity can be reduced by pressures exerted by a population, at least in some circumstances, and when this occurs there is a situation that is equivalent to a population reducing its own resources. I have not, at this juncture, specified exactly what constitutes the carrying capacity but in practice it nearly always means the availability of food.

The distribution of populations

If the distribution of a species of animal is plotted on a map of Africa a distinct pattern emerges. If the exercise is repeated for several different species, it soon becomes apparent that some have a wide and continuous distribution, perhaps paralleling a vegetational or climatic zone, while others are extremely restricted. For many invertebrates, including a majority of the species of insects, there is so little information available that when the known distribution is plotted there is no more than a few dots on the map. Nevertheless this is better than nothing and represents a beginning of an understanding of the distribution of populations.

The turacos belong to a family of birds confined to Africa. There are 18 species and most of them have a restricted distribution. Figure 3.1 shows the distribution of two species, the Ruwenzori turaco, *Tauraco johnstoni*, and the great blue turaco, *Corytheola cristata*. *Tauraco johnstoni* occurs in: highland forest on Ruwenzori; the highlands north-east of Lake Edward; the Kivu volcanoes; the highlands north-west of Lake Tanganyika; and nowhere else. In contrast *Corytheola cristata* ranges from Guinea and Fernando Póo (now Macias Ngnema) in the west right across equatorial Africa to western Kenya. It occurs as far north as 9°40′N. in

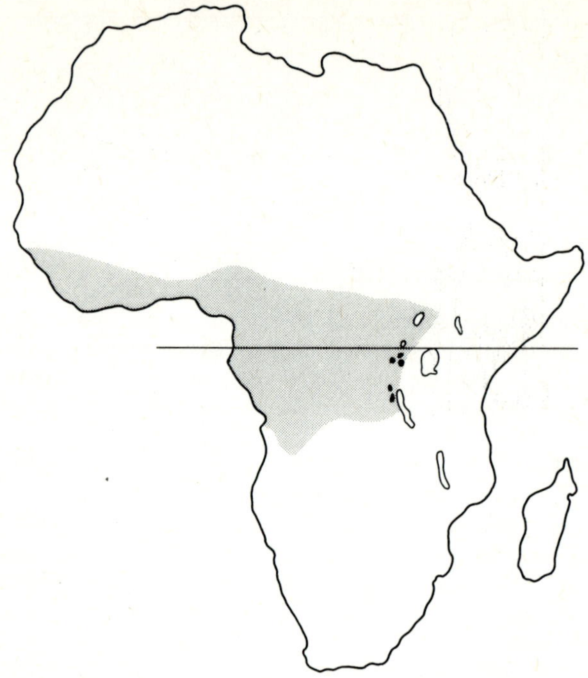

Fig. 3.1 The distribution of two species of turaco. The small isolated black areas indicate the distribution of *Tauraco johnstoni*, the large shaded area the distribution of *Corytheola cristata*.

Nigeria and south to about 11°S. in Angola. It is a bird of lowland forest and forest edge, and also occurs in savanna where there are plenty of trees. The wide distribution of this species can be compared with the restricted distribution of *Tauraco johnstoni*: one species is found throughout the lowland equatorial belt while the other is confined to highland forest. It is clear that the distribution of these two types of habitat determines the distribution of the birds; but this is a descriptive statement and tells nothing of the means by which these distributions are maintained or why one species does not spread into the area occupied by the other.

Figure 3.2 shows the distribution of the gorilla, *Gorilla gorilla*, in central Africa. As can be seen, the species occurs in many isolated or relatively isolated populations. Rivers form a major barrier since gorillas do not normally cross them. The gorilla is a stem- and pith-eater and its distribution is to some extent limited by the availability of suitable food. In some areas it is persecuted by man and its range has decreased, but in others it has increased as it exploits crops grown by man. There is at the moment no real danger of gorillas becoming extinct, but their numbers may decline rapidly if human numbers continue to rise in the areas where they occur. Unfortunately there is still a demand for gorillas for European and American zoos. The only way a baby can be taken alive is by killing the mother, and I hope that before long there will be a complete ban on the export and import of gorillas.

Figure 3.3 shows the present distribution of the waterbuck, *Kobus ellipsiprymnus*, in Kenya. It occurs chiefly along river valleys and in low-lying country. This pattern of distribution is found in many other animals that live in seasonally dry parts of tropical Africa. West of the Rift Valley the place of *K. ellipsiprymnus* is taken by a similar species *K. defassa* and it too is closely

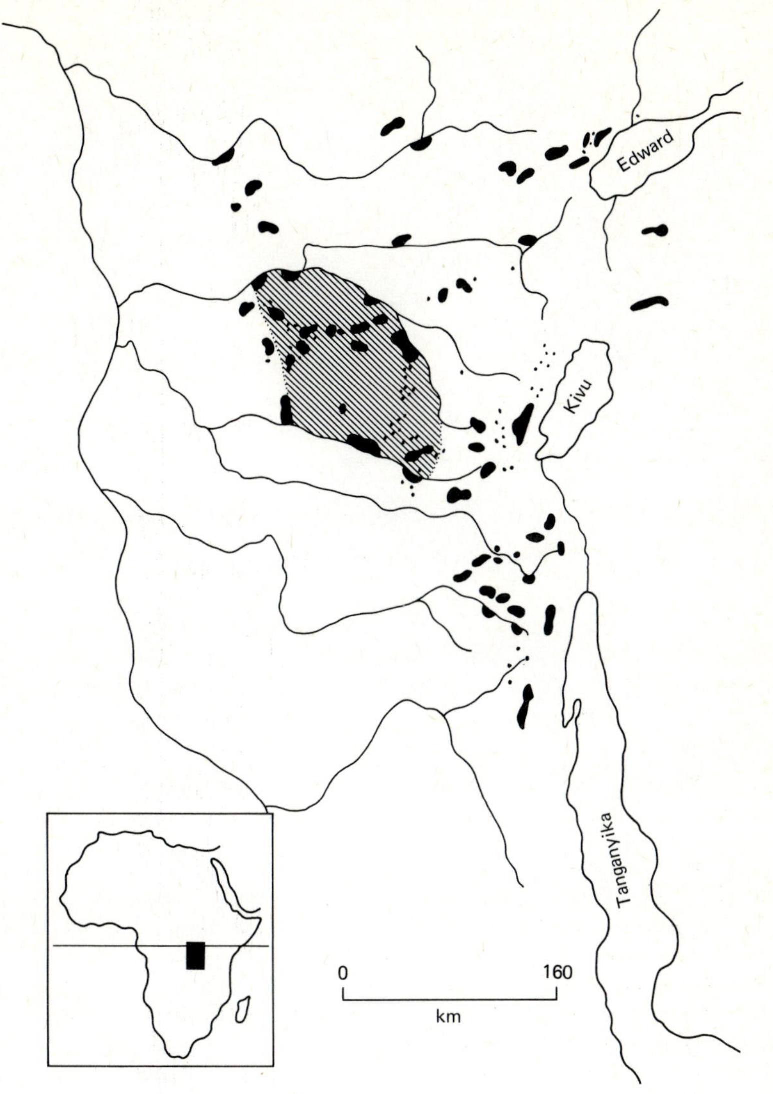

Fig. 3.2 The distribution of the mountain gorilla, *Gorilla gorilla*, in central Africa. The black areas are isolated populations and the shaded area indicates a probably continuous distribution. (From Schaller, 1963.)

associated with river valleys. The ranges of the two species overlap where they meet.

Figure 3.4 shows the present distribution of: the elephant, *Loxodonta africana*; buffalo, *Syncerus caffer*; topi, *Damaliscus korrigum*; and eland, *Taurotragus oryx*, in Uganda. Judged by distribution the general ecological requirements of the elephant and the buffalo are similar, as are those of the topi and eland. The elephant and buffalo occur where there are at least some trees and in particular where seasonal drought is less marked; the topi and the eland can tolerate more open and drier country. Figure 3.5 shows the areas of Uganda where the mean annual rainfall is above and below 100 cm. The association of the elephant and the buffalo with areas of higher

Fig. 3.3 The distribution of the waterbuck, *Kobus ellipsiprymnus*, in Kenya. Major mountains are shown and the dashed lines indicate the position of the Rift Valley. (From Stewart and Stewart, 1963.)

rainfall and the topi and the eland with areas of lower rainfall is conspicuous, except near Lake Victoria in southern Uganda which is heavily settled by man and is hence unsuitable for large mammals. But although such an association between rainfall and distribution of species of large mammals exists, exactly what ecological factors limit the distribution in this way are not known.

Figures 3.1 to 3.4 show that species have characteristic distributions. Some species are obviously restricted by features of the environment such as mountains, rivers, forests, and the amount of rainfall. But so far the problem of numbers has not been considered: these distribution maps simply show whether a species is present or absent in a particular area.

The dispersion of individuals in populations

The word dispersion is here used to denote not only where a particular animal is found but also how many there are in a limited area. Animals are not randomly spaced: there is generally a well-defined pattern in the frequency in a given area. This frequency-pattern is what is meant by dispersion. The simplest example of non-random dispersion of a species is in man. Man is most abundant, and towns and villages are larger, in areas where resources and means of communication are best. The availability of food, shelter, the relative freedom from predators and parasites, and a favourable climate are the kinds of environmental factors that determine the dispersion of individuals in a population. It is, however, by no means easy to provide examples of dispersion chiefly as a result of the difficulties in counting that arise because of the mobility of most species of animals. There is much more information about dispersion in plants which being

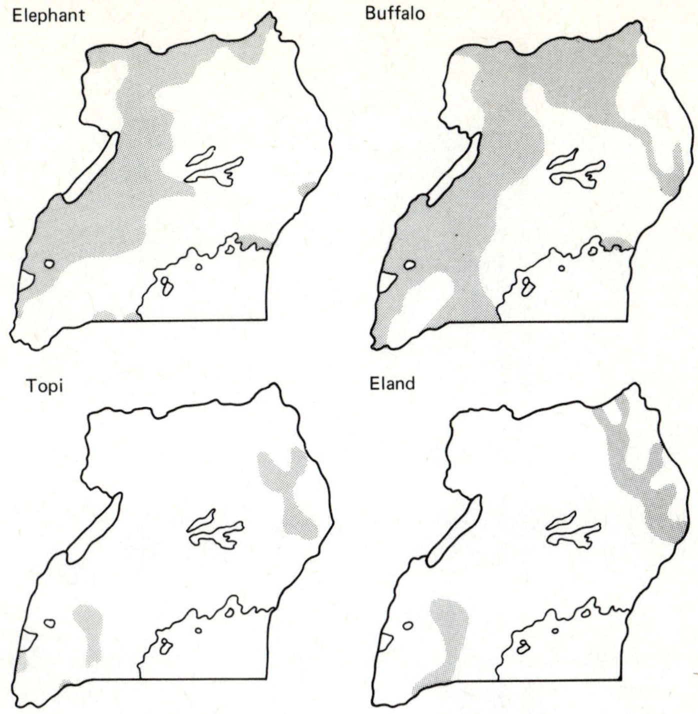

Fig. 3.4 The distribution of: the elephant, *Loxodonta africana*; buffalo, *Syncerus caffer*; topi, *Damaliscus korrigum*; and eland, *Taurotragus oryx*, in Uganda.

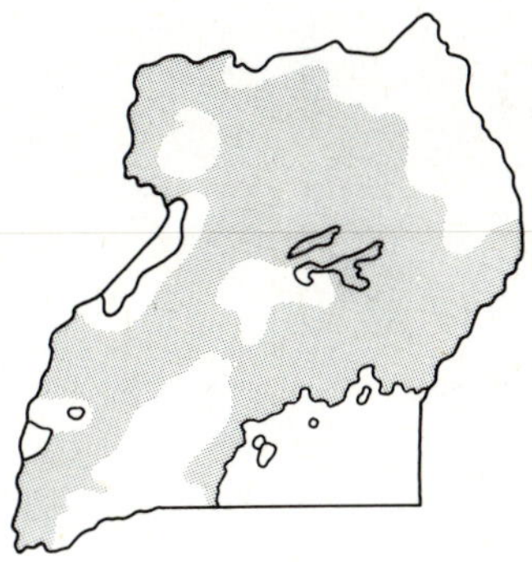

Fig. 3.5 Uganda: mean annual rainfall. Shaded area – over 100 cm; unshaded area – under 100 cm.

immobile are easier to count. Two examples of dispersion in populations of tropical African animals, one a land snail, the other a bird, will be considered.

In East Africa the herbivorous land snail, *Limicolaria martensiana*, occurs in well-defined populations separated from each other by ecological and geographical barriers. The snails are 3 to 4 cm long when they are adult and in places where they occur they are conspicuous. Table 3.1

Table 3.1 *Dispersion of the land snail,* (Limicolaria martensiana) *in Population A at Kampala between 14 February and 13 March 1963. Each figure represents the number of living snails in a 1-metre quadrat taken in the south-west corner of a 20-m square. The solid lines show the approximate position of the 25-ft* (c. *7·6 m*) *contours. The dotted line indicates the position and direction of flow of a small stream.*

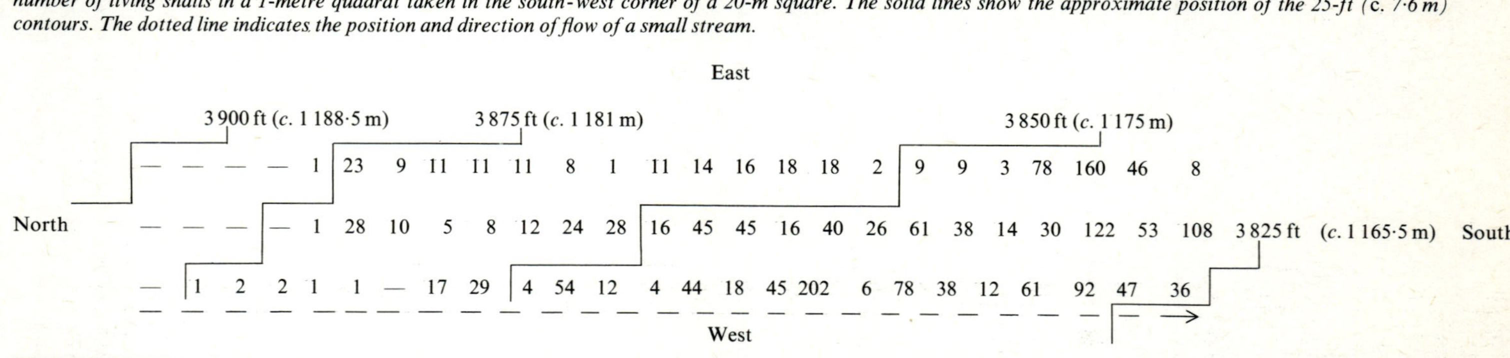

East

3 900 ft (*c.* 1 188·5 m) — 3 875 ft (*c.* 1 181 m) — 3 850 ft (*c.* 1 175 m) — 3 825 ft (*c.* 1 165·5 m)

North → South

—	—	—	—	1	23	9	11	11	11	8	1	11	14	16	18	18	2	9	9	3	78	160	46	8
—	—	—	—	1	28	10	5	8	12	24	28	16	45	45	16	40	26	61	38	14	30	122	53	108
—	1	2	2	1	1	—	17	29	4	54	12	4	44	18	45	202	6	78	38	12	61	92	47	36

West

shows the dispersion of snails in Population A at Kampala, Uganda. The figures shown are the number of snails in a 1-m frame quadrat taken at regular intervals throughout the area occupied by the population. The population extends over an area of 30 000 m^2. The number of snails in each quadrat varies from 0 to 202. It is clear that the dispersion of snails in the population is not even and it can be shown to be non-random. Snails become progressively more frequent from north to south and also from east to west. The 25-ft (*c.* 7·6 metre) contours are shown diagrammatically in Table 3.1. Between the lowest two contours there are 46·7 snails per quadrat, while between the middle two contours there are 10·7 snails per quadrat. Above the highest contour snails are rare: only one out of nine quadrats contained a snail. The slope of the land, and particularly the presence of a small stream on the west side of the area, results in the south-west corner being damp and the north-east corner relatively dry. The gradient in wetness from north-east to south-west accounts for the dispersion of the snails in this population. Other factors may also be involved, such as the availability of food, which in turn may be correlated with moisture.

In another population (Population B) of the same species the dispersion of the snails depends to some extent upon the occurrence of a particular plant. Population B occupies an area of 3 450 m^2; there is no noticeable slope and evidently no moisture gradient. The population was sampled in the same way as Population A. The results for two series of samples on two dates are shown in Table 3.2. The pattern of dispersion is similar, indicating stability. The snails are

Table 3.2 *Dispersion of the land snail,* Limicolaria martensiana, *in Population B at Kampala on 10 April 1963 and 5 November 1963. Each figure represents the number of living snails in a 1-metre quadrat taken in the north-east corner of a 20-m square.*

10 April 1963

East

	1	3	19	27	8	12	
North	2	24	22	39	115	17	South
	—	4	7	16	32	28	

West

The mean density inside the solid line is 30·5 per m^2; outside it is 8·9 per m^2.

5 November 1963

East

	—	8	11	27	16	7	
North	5	33	33	58	63	4	South
	1	22	28	25	52	48	

West

The mean density inside the solid line is 35·9 per m^2; outside it is 10·3 per m^2.

relatively rare in the north and north-west and rather rare in the south-east; they are concentrated about the middle. The solid lines in Table 3.2 indicate the area in which a favoured food-plant, *Bryophyllum pinnatum*, covers more than 50 per cent of the ground. The association of increased numbers of snails with a greater amount of *Bryophyllum pinnatum* is not spectacular (the snails utilize a wide variety of other plants) but dispersion is clearly related to the abundance of this one food-plant.

Dispersion within these two populations of *Limicolaria martensiana* is neither random nor even but is related to environmental factors, two of which (there are undoubtedly others) are a moisture gradient and the abundance of a particular plant. Land snails are slow-moving animals and presumably patterns of dispersion such as these are quite rigidly fixed and are maintained for long periods.

The fish eagle, *Cuncuma vocifer* (Fig. 3.6), is a large and conspicuous bird in tropical Africa. It

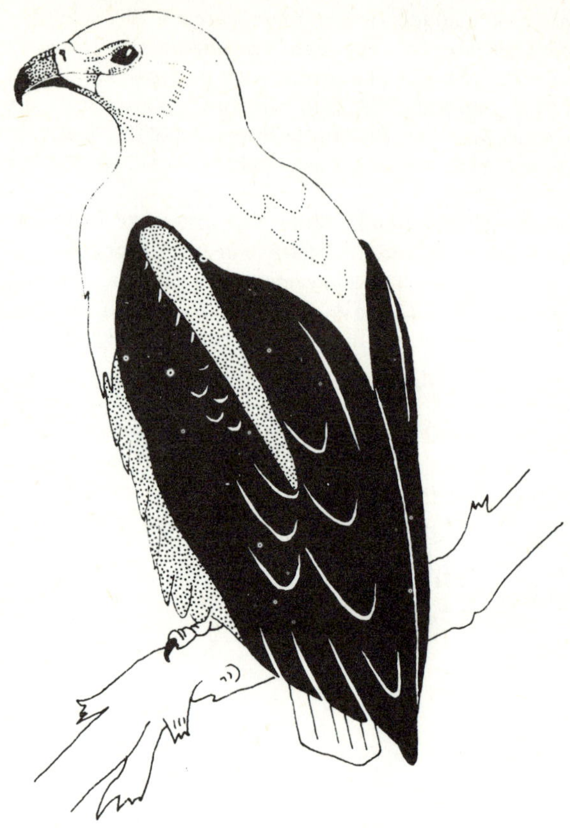

Fig. 3.6 The fish eagle, *Cuncuma vocifer*.

feeds on lake and river fish, catching its own prey and scavenging around fishing villages. Fish eagles tend to be spaced at intervals along the shores of lakes and rivers. In 1962 a census was made of the fish eagles along the eastern shore of Lake Albert (Sessekou Mobutu) (Green, 1964). About 209 km of shore was examined and 167 fish eagles were counted; this gives an average of 0·8 eagles per km. The dispersion of the fish eagles and the position of fishing villages is shown in Fig. 3.7. The dispersion of the eagles appears to be determined by three factors. First, fish eagles use conspicuous perches with a wide view of the lake. These are usually dead trees at the edge of the water. Fish eagles are relatively rare where perches are scarce, as along the southern section of the shore where the escarpment of the Rift Valley plunges steeply into the Lake. Secondly, they are frequent where the water is shallow and fish more available. Where the Nile enters the northern part of the Lake there are extensive papyrus swamps and here fish eagles are numerous. Thirdly, numbers are higher around fishing villages, partly because the birds are able to scavenge and partly because, like fishermen, eagles tend to concentrate where fish are most available.

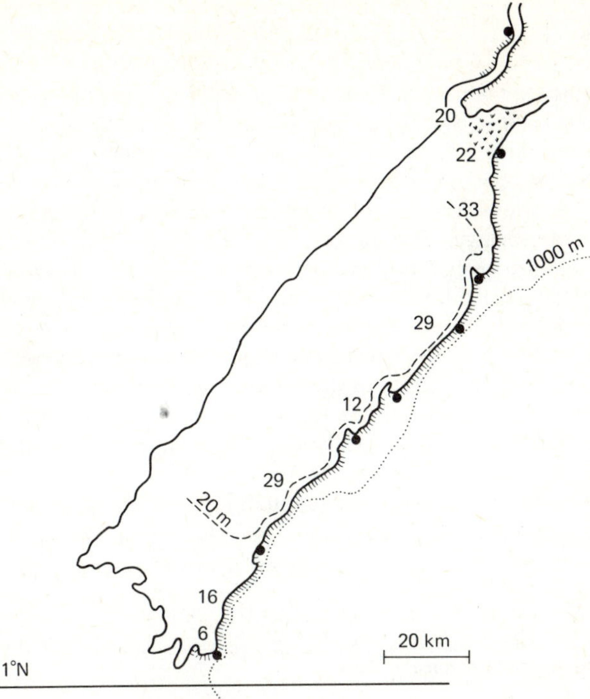

Fig. 3.7 The dispersion of the fish eagle, *Cuncuma vocifer*, along the eastern shore of Lake Albert, Uganda. The 1 000-m contour marking the edge of the Western Rift Valley and the 20-m depth line are shown. Solid dots mark the position of fishing villages. Where the Nile enters and leaves the Lake at its northern end, there are extensive papyrus swamps. The shoreline censused is shown by hatching. (From Green, 1964.)

Hence the dispersion of the fish eagle on the shores of Lake Albert is not random or even but is determined in a variety of ways. It is important to note that suitable food must not only be abundant but that it should also be available. It appears that fish eagles and fishermen share an interest: both are more concentrated where resources are more available.

Dispersion and behaviour

There are in many species behavioural mechanisms by which individuals become spaced out. Many birds and fish, some mammals and insects (especially butterflies) and some other animals isolate themselves from others of their species at the time of breeding. Many of these species are territorial. A territory is an area that is defended against other members of the same species. One outcome of territorial behaviour is that individuals acquire for themselves a private space within which they can live and breed. In most birds and fish territories are defended only in the breeding season; outside the breeding season they may be social and gregarious. The result of territorial behaviour is that animals are spaced out and hence resources, which may include food and a suitable place to breed, are better utilized. The establishment of a territory also increases the probability of obtaining a mate: males of many species make themselves conspicuous by their territorial behaviour and so attract suitable females into the territory.

In birds, territorial behaviour is somewhat less conspicuous in tropical than in temperate species. This is because many tropical species (especially those that live in forests) remain in the same area all their lives whereas most temperate species breed in areas other than those in which they spend the rest of the year. Temperate species thus have to re-establish themselves in their breeding territories each time they breed; this is not so in a majority of tropical species.

One of the common antelopes (the kob, *Adenota kob*) is in parts of Uganda territorial all the year round (Buechner, 1961). Small territories are defended within a larger area of concentrated territorial activity; this area is surrounded by a zone of more widely spaced territories. Females may enter the defended area at any time of the year for the purpose of mating. This particular type of territorial behaviour does not appear to have been observed in other tropical mammals and seems associated with high population density.

Social species may also be dispersed in such a way as to utilize the available resources more effectively. Colonies of social birds such as weavers, and colonies of termites and ants, are spaced out in such a way as to ensure effective utilization of food and other resources. Some of the species of weavers breed in colonies in villages occupied by people or in trees occupied by large predators such as eagles; this pattern of dispersion would appear to afford the colony some degree of protection from enemies. In some animals, for example social wasps and hunting dogs, there is a division of labour within the social group. The wasps have a worker caste that assist in the feeding of the larvae and in the care of the nest. In the hunting dog, *Lycaon pictus*, all the adults are equally qualified for hunting and for guarding and feeding the young, but at any one time only a few of the adults take part in a hunt; the others remain behind and take care of the young. The meat is brought back by the hunters and is shared among the guards and the young (Kühme, 1965).

Population size and population density

Finding out the distribution of species and the dispersion of individuals and populations is the first step in any attempt to evaluate the factors that limit the size of populations. The next step is to find out the size and density of populations.

As pointed out in Chapter 2 some tropical African species are common and many are relatively rare, but apart from the large mammals of the East African savanna and a large variety of species of invertebrates and small vertebrates in the savanna of the Ivory Coast, rather few estimates of population size and density of tropical African animals are available. This is chiefly because of the great practical difficulties involved in obtaining such estimates, although these can be overcome as has been demonstrated in the Ivory Coast. Two general methods are available: direct counting and sampling. Direct counting is effective if the animals are large and conspicuous (the large mammals of the savanna can be counted with relative ease) or if the animals are sedentary. Sampling can be effective for animals that are relatively small and that do not move around too much: populations of land snails, soil invertebrates, and certain colony-forming species of butterflies can be readily sampled and reliable estimates of population size and density can be obtained. Most populations of flying animals (bats, birds, and adult insects) are by no means easy to estimate and nocturnal and aquatic species present special problems. I do not intend to discuss here the many sampling methods available for different groups of animals, neither shall I discuss the difficulties that can be encountered. Such information is given in some of the numerous ecology text-books that are nowadays available – in the examples cited below the author has in most instances described in detail the procedures by which estimates were obtained.

Bourlière (1963) has summarized some of the estimates of population size and density of large mammals, chiefly herbivores, that have been made in the East and Central African savanna and to a lesser extent elsewhere. Some of the large herbivores may have very high population densities especially in areas that have been set aside as national parks.

Table 3.3 shows the numbers of some large herbivorous mammals on the Rwindi-Rutshuru

Table 3.3 *Numbers, population density, and biomass of some large herbivorous mammals on the Rwindi-Rutshuru Plain, Eastern Zaïre. The figures are the means of six counts made in 1959. (From Bourlière, 1963.)*

Species	*Average adult weight, in kg*	*Number in 600 km²*	*Density per km²*	*Biomass per km² in kg*
Elephant, *Loxodonta africana*	3 000	1 026	1·7	5 100
Hippopotamus, *Hippopotamus amphibius*	1 400	4 800	8	11 200
Buffalo, *Syncerus caffer*	500	7 402	12·3	6 150
Topi, *Damaliscus korrigum*	130	1 199	2·0	260
Waterbuck, *Kobus defassa*	150	760	1·2	195
Uganda kob, *Adenota kob*	70	4 976	8·3	581
Warthog, *Phacochoerus aethiopicus*	70	603	1	70

Note: About five additional species of large herbivores occur, but they are relatively rare. No figures are available for small mammals, primates, and members of the order Carnivora.

Plain of Eastern Zaïre; the average weight of each species and the biomass per km^2 is also given. Biomass is calculated by multiplying the average weight by the density: it is a useful measure of the 'amount' of each species in an area. The combined biomass of large herbivores on the Rwindi-Rutshuru Plain, 22 556 kg per km^2, is probably the highest figure for any natural area in the world. About 69 per cent of the biomass is made up of two species, the elephant and the hippopotamus. Many comparable estimates are available for other areas in East and Central Africa.

Table 3.4 gives some estimates of numbers, density, and biomass in an area of West African forest. The combined biomass of the three species of ungulates is only 5·6 kg per km^2. Diurnal primates (monkeys and the chimpanzee) are the most conspicuous mammals of the West African forests, but the seven species listed in Table 3.4 comprise only 66·6 kg per km^2. Some of the forests of Zaïre and Uganda support large populations of elephant and buffalo but no reliable estimates of numbers are available. In general, large mammals are much scarcer in the forest than in the savanna chiefly because they are largely dependent on grass.

Table 3.4 *Numbers, population density, and biomass of ungulates and primates in the Tano Nimri Forest Reserve, Ghana, in 1954. (From Bourlière, 1963.)*

Species	*Average adult weight, in in kg*	*Number in 250 km²*	*Density per km²*	*Biomass in kg per km²*
Duiker, *Philantomba maxwelli*	8	79	0·31	2·48
Duiker, *Cephalophus dorsalis*	20	38	0·15	3
Antelope, *Neotragus pygmaeus*	4	7	0·03	0·12
Black colobus, *Colobus polykomos*	10	916	3·6	36
Red colobus, *Colobus badius*	8	621	2·4	19
Diana monkey, *Cercopithecus diana*	5	144	0·57	2·8
Mona monkey, *Cercopithecus mona*	5	127	0·50	2·5
Mangabey, *Cercocebus torquatus*	8	83	0·33	2·6
Olive colobus, *Colobus verus*	4	5	0·02	0·08
Chimpanzee, *Pan troglodytes*	40	22	0·09	3·6

Note: Neotragus pygmaeus, the royal antelope, is the smallest ungulate in the world (Fig. 3.8). The average weight given by Bourlière is twice that of a captive individual that we kept for a time in Sierra Leone.

Fig. 3.8 The royal antelope, *Neotragus pygmaeus*, the smallest antelope in the world, smaller than a rabbit. It occurs in West African forests, where it feeds on flowering plants growing on the forest floor. Photographed in Sierra Leone.

If, however, all animals including insects and other invertebrates were considered, the biomass of forest animals would be considerably higher though certainly never as high as the savanna. On the other hand there is little doubt that the forest supports a much greater number of species than the savanna but no overall comparison is available. Certainly the African savanna is capable of supporting a much greater biomass of animals than any other natural environment on earth.

Some of the drier areas in East Africa support large numbers of birds of prey. Up to 12 species of eagle and eagle-like birds can occur together. Table 3.5 shows the results of a census of large birds of prey made during 3 consecutive years in the Embu District of Kenya (Brown, 1955). The area includes rocky hills, heavily bushed river valleys, and wooded savanna. Apart from the fish eagle (which feeds upon fish) and the secretary bird (which feeds upon grasshoppers as well as vertebrate prey) all species are predators of: small- or medium-sized mammals; large passerine birds; game birds (including domestic chickens); lizards; and snakes. A large number of prey species has been recorded as being eaten by the eagles but possibly they are collectively mainly dependent on a few species, such as hyraxes, game birds, and certain snakes and lizards. As shown in Table 3.5, with the exception of the secretary bird numbers remained stable in each of the 3 years; this suggests a close adjustment between the number of eagles and their available prey.

By far the most detailed estimates of population density and biomass (though not usually

Table 3.5 *Number of pairs of large birds of prey in 146 miles² (c. 378 km²) in the Embu District of Kenya. (From Brown, 1955.)*

	Number of pairs		
	1950	1951	1952
Secretary bird, *Sagittarius serpentarius*	4–5	1	—
Verreaux's eagle, *Aquila verreauxii*	1	1	1
Wahlberg's eagle, *Aquila wahlbergi*	8	9	11
African hawk-eagle, *Hieraaetus spilogaster*	2	2	2
Ayres' hawk-eagle, *Hieraaetus ayresi*	1	1	1
Martial eagle, *Polemaetus bellicosus*	3	3	3
Crowned hawk-eagle, *Stephanoaetus coronatus*	1	1	1
Long-crested hawk-eagle, *Lophoaetus occipitalis*	1	1	1
Bateleur, *Terathopius ecaudatus*	2	2	2
Brown harrier-eagle, *Circaetus cinereus*	2	2	1
Black-breasted harrier-eagle, *Circaetus pectoralis*	1	—	—
Fish eagle, *Cuncuma vocifer*	1–2	2	2

Not every pair nested successfully.

population size) are for a whole variety of animals in the wooded savanna (*préforestière*) at Lamto in the Ivory Coast. Report after report has come from Lamto and more is known about the fauna of this region than anywhere else in Africa; space permits mention of only a few of these studies.

More than 400 species of birds have been found in the wooded savanna and adjacent gallery forest at Lamto and the population density of many of these has been calculated (Thiollay, 1973). Many species occur at a density of less than 1 per 50 ha and the commonest species, the yellow-mantled whydah, *Euplectes macrourus*, at a density of 75 per 50 ha which is not especially common. Most of the species are more abundant at the height of the wet season (July and August) than in the dry season (January–March). The green pigeon, *Treron australis*, is seven times as common in the wet as in the dry season but there are reversals of this tendency as with the broad-billed roller. *Eurystomus glaucurus*, present only in the dry season and presumably a migrant from drier savanna to the north.

Casual observations suggest that lizards, skinks, and geckos are abundant virtually everywhere in tropical Africa, but how common is each species? There is little information. At Lamto one of the common skinks. *Mabuya büttneri* occurs at densities of between 30 and 60 to the hectare (Barbault, 1971), but in this skink, as in other small reptiles, there is no information on population size, largely because of difficulties in defining the geographical and ecological limits of the population.

Casual observations suggest also that earthworms, although scarce in many areas, may in places reach the level of abundance that they do in temperate pastures. But again there is little information. One species, *Millsonia anomala*, which grows to a length of 25 cm and is therefore a very large worm, occurs at a density of about 18 to the square metre at Lamto (Lavelle, 1971). Somewhat similar estimates of density and biomass have been made for various smaller species in Nigeria and Uganda but with one or two exceptions earthworms do not appear to be as abundant in tropical Africa as in most temperate areas.

In the land snail, *Limicolaria martensiana*, mentioned earlier in this chapter, it has in some instances been possible to define precisely the limits of the population and to arrive at estimates of total population size. Thus at Kampala, Population A occurs over a rectangular area of 30 000 m². Roads, buildings, and open grassland surround the area and the population is thus completely isolated from others in the vicinity. In 1963, when the population was first sampled with a frame quadrat it was found that the snails occurred at a mean density of 26·4 to the square metre which gives an estimated population size of 792 000. From 1964 onwards the Kampala

City authorities decided that the vegetation of the area should be cut down to, and maintained at, ground level and in 1966 when the population was again sampled the density was only 2·1 to the square metre, 10 per cent of the 1963 estimate. This indicates a fall in population size to about 84 000 and, in this instance, the reduction was undoubtedly caused by the removal of cover. Exactly how the snails were reduced in numbers is a matter for speculation: possibilities include food shortage, exposure to the sun and generally drier conditions, and increased predation by birds – all of which could be effected by removal of most of the cover.

With land snails, all individuals, including the very young, can be sampled by using a frame quadrat, but with most insects it is only the adults that can be sampled because the larvae and pupae are dispersed and hidden in such a way that the use of reliable sampling techniques is prevented. A few butterflies, like land snails, form discrete populations the boundaries of which can be identified and defined. One of these is *Acraea encedon* (Fig. 3.9), a slow-flying species that

Fig. 3.9 The butterfly *Acraea encedon.* In some places its population size varies on a seasonal basis while in others the low frequency of males generates irregular fluctuations in numbers.

tends to form small colonies confined to relatively small areas throughout tropical Africa. It is possible, in this species, to catch and mark on the wing with a spot of coloured ink every individual flying on a particular day, an operation that provides information on population size and fluctuations in population size. At one locality in Sierra Leone the adult population was determined once a week for 50 consecutive weeks and once every 2 weeks for a further 32 weeks (Owen, 1971). During this period the population size fluctuated between 0 and 136; the highest numbers occurred at the end of the wet season and early part of the dry season, while the butterfly was scarce or absent at the end of the dry season and beginning of the wet season. This cycle in numbers, known to be repeated every year in this population, seems associated with changes in the availability of the larval food-plant (*Commelina* spp.) which tends to die back in the dry season and to grow vigorously when it is wet. In other populations of *Acraea encedon*, particularly in Uganda which has a less conspicuously seasonal climate than Sierra Leone, fluctuations in population size are correlated with small changes in the frequency of males. Since these, in any case, are always rare many females remain unmated and contribute nothing to succeeding generations. Hence in *Acraea encedon* the factors that limit population size differ from place to place – a discovery that, were there sufficient information, would probably apply to many other animal populations.

Apart from the large mammals which nowadays are mainly confined to national parks, and a few species of invertebrates like *Limicolaria martensiana* and *Acraea encedon*, there is remarkably little information about the size of populations of animals in tropical Africa. The relative abundance of species, especially species with similar habits, is not difficult to obtain even in a habitat like a garden. For example, using the simple technique of capture, mark, and release, I obtained, during a year, 423 individuals of 12 species of swallowtail butterflies in a garden in Sierra Leone. Most of them entered the garden to feed from flowers; a few came to lay eggs. Three-quarters of the individuals belonged to one species, *Papilio demodocus*, a further 4 species occurred 10 or more times each, while 3 species appeared once only. This kind of sampling tells nothing about the size and density of populations, but if repeated in a variety of habitats and localities it would provide information that would begin to suggest reasons why, for example, a butterfly may be rare in one place and common in another. Routine sampling and counting of this sort could be achieved by teams of students from schools and colleges as class exercises and would provide valuable experience in practical ecology as well as contributing knowledge about the African environment.

The frequency of parasites on their hosts

Here again is a problem that given the appropriate choice of material offers scope for a class exercise. There are parasites on or in all animals and it is often easy to find evidence of their occurrence. The host provides the food resources (and indeed the shelter) for the parasite; by definition parasites do not normally kill their hosts, but they occasionally have severe effects and may sometimes cause death. All individual parasites of a species on a host constitute a population (perhaps better called a 'mini-population') and since each host is usually physically isolated from others of its species it is possible to compare the rates of parasitization on different individual hosts. It might be expected that given a population of hosts of roughly comparable size and age there would be, on average, about the same number of parasites on each of them. But this expectation turns out to be the exception rather than the rule as shown in the following examples.

If a sample of hosts is examined and the frequency of a particular species of parasite on each individual is recorded, it is often found that most of the parasites are associated with relatively few of the hosts. Table 3.6 shows the numbers of a parasitic crustacean, *Argulus africanus*, found

Table 3.6 *The frequency of a parasitic crustacean on a fish. (From Williams, 1964.)*

Parasites per fish	*Number of fish*	*Parasites per fish*	*Number of fish*
0	154	14	3
1	70	15	2
2	25	16	1
3	22	17	3
4	6	18	4
5	8	19	3
6	4	20	3
7	8	21	2
8	3	22	2
9	1	24	2
11	4	28	2
12	7	30	1
13	1	34	1

on a large sample of the cichlid fish, *Tilapia esculenta*, in a tropical African lake. There is a mean of 3·18 parasites per host, but over half the parasites are on only 8 per cent of the fish. Thus a few fish are very heavily parasitized and most fish are lightly affected. This kind of distribution of parasites on hosts can be shown to be neither random nor even. Similarly, in a sample of 560 specimens of the Lake Malawi fish, *Bagrus meridionalis*, 333 were infected with the parasitic crustacean, *Lernaea bagri*, and again most of the fish were lightly infected and a few very heavily infected (Fryer, 1956).

The same kind of frequency-distribution has been found in that curious parasite, *Oculotrema hippopotami*, that lives on the eyes of the hippopotamus (see page 41). In a sample of 42 hippopotamuses from Ruwenzori National Park, Uganda, 38 were infected with the parasite. There was a mean of 8·3 parasites per hippopotamus, but 52 per cent of all the parasites collected were found on only 6 of the 42 hippopotamuses (Thurston, 1968).

Indeed almost exactly the same kind of frequency-distribution occurs again and again in a wide variety of parasite-host relationships. If, for example, a sample of mature maize or sorghum stems is collected, each stem split, and the frequency of pith-feeding moth larvae recorded, the same frequency-distribution is found: most stems will contain a few larvae but most larvae will occur in a few stems.

It appears that even though individual hosts are apparently similar in the resources they offer to a parasite population there are striking inequalities in the level of infection in different individuals. There seems to be no satisfactory explanation of this non-random occurrence of parasites in hosts. It is, however, reminiscent of the frequency of occurrence of different species that emerges when a large sample of taxonomic group is taken: most species are relatively rare and a few are common (Ch. 2).

Even though not understood the phenomenon has important medical implications. Thus the frequency of *Schistosoma* in man shows a virtually identical pattern: in many parts of tropical Africa most people suffer from schistosomiasis but few suffer seriously from the disease; indeed serious effects from the parasites occur only when there are large numbers of them in a single individual. This, I believe, has led to the considerable controversy as to what should be done about the disease, possibly explaining why research on schistosomiasis has tended to lag behind that on other common tropical diseases, such as malaria.

Factors limiting population growth

The population size and population density of a species vary in different environments and species differ markedly in abundance. Each environment is characterized by an assemblage of rare or relatively rare species and a few common ones. With this information it is now necessary to consider the ways in which populations are regulated in size, and it is at this point that I become rather more theoretical.

Populations may be regulated in two ways: by density-*dependent* or by density-*independent* events. Density-dependent events are environmental factors that operate differently at different population densities. In particular, the death rate increases with increasing population density and decreases with decreasing population density. The availability of food and the pressure exerted by predators seem to act density-dependently on most populations, but this does not mean that food availability and predator pressure always act in this way even in the same population. Density-dependent events are not easy to detect in the field although there is some direct and much suggestive evidence of their action. Density-independent events are environmental factors whose effect is not related to the density of the population. Density-independent mortality seems to occur when there is a sudden environmental change that affects all members of the population equally. A grass fire could kill all members of a snail or insect population no matter what the population density. A rise in water level could also kill density-independently. Figure 3.10 shows the effect of the recent rise in the level of Lake Victoria: trees

Fig. 3.10 Forest trees that have been killed by the rising level of the water in Lake Victoria. This and other tropical African lakes are subject to irregular fluctuations in the level of the water, partly because of variations in rainfall and partly because of human activities such as dam building and drainage projects. (Photograph by C. H. F. Rowell.)

around the edge have been killed by the water, and this mortality is independent of the density of the trees. It may be assumed that sedentary animals on the living trees would also die density-independently.

Density-dependent events tend to maintain a population at relative stability at a level that depends on the carrying capacity of the environment. Fluctuations in population size are small compared to what is potentially possible because whenever numbers tend to become too high the death rate increases. The size and density of a given population depends on the resources available to it; such resources include the availability of food, shelter, breeding places, and so on, although for a majority of populations it appears that food is the crucial factor. Density-independent events, on the other hand, tend to produce large fluctuations in numbers and occasional extermination of the population. Most ecologists agree that density-dependent events operate in nearly all populations and that density-independent events are less important. It is likely, however, that many populations are at least occasionally subject to density-independent regulation. The effects of density on death rates have been demonstrated in laboratory populations of insects and micro-organisms and occasionally in field situations; but in the main, and especially so far as populations in tropical Africa are concerned, the evidence must be considered as essentially circumstantial.

If it is accepted that most populations are limited primarily by density-dependent events it follows that as numbers tend to rise there must be competition for resources among individuals. Competition is likely whenever animals share the same controlling factors; that is to say, when there are not enough resources available, some individuals die while others survive. Resources in this context can be taken to mean not only food but also, for example, suitable places to hide from predators. There is much indirect and some direct evidence that competition for food is the chief limiting factor for many animal populations. It is likely that most populations of predatory animals (including the many species of predatory insects) are food-limited and hence there is competition among individuals for the available prey. On the other hand it is easy to form the impression that herbivorous animals feeding on living vegetation have an unlimited supply of food; but appearances are deceptive for much of the green vegetation is inedible to many animals: leaves are often tough and unpalatable and an extraordinary variety of plants elaborate chemical compounds that are toxic to all but a few of the herbivores. Nevertheless the observation that very little green vegetation, especially in forests, is consumed has led some ecologists to propose that plant-feeders, as a group, are limited in numbers by their predators rather than by the availability of food.

In gardens throughout tropical Africa citrus trees are planted or are allowed to grow freely. Swallowtail butterflies of several species lay their eggs on new leaves: the larvae cannot eat the old tough leaves. Young swallowtail larvae similar and smaller in size to that shown in Fig. 3.11 therefore depend on the availability of soft, new leaves; most of the leaves on the citrus are unacceptable. But as the larvae grow (Fig. 3.12) they are able to eat older leaves, and yet despite the abundance of food hardly ever attain full growth. They are subject to intensive predation by mantids, ants, and other insects. If they are brought indoors and reared on leaves gathered from the tree they usually attain full size. Hence it seems that the numbers of these swallowtails may be limited as young larvae by the availability of soft leaves but that later they are limited by predators.

The savanna areas of Africa experience seasonal drought and although most of the time there may be adequate food for the large herbivorous mammals there may also be times of food scarcity. In some of the savanna areas of East Africa that have been set aside as national parks, elephants have become extremely abundant. The 'elephant problem', as it is called, has generated much disagreement among ecologists. The 'problem' arises because under protection and in relatively confined areas from which the animals can no longer undertake seasonal migrations, numbers build up to an extent that the vegetation is seriously damaged. This in turn adversely affects the abundance and diversity of other species of herbivorous mammals (and

Fig. 3.11 A young larva of the swallowtail butterfly, *Papilio demodocus,* feeding on citrus leaves.

their predators) which, of course, are not only of scientific interest but a significant source of revenue as tourist attractions.

In the vicinity of Murchison Falls National Park in Uganda, 88 per cent of the total food consumed by elephants is grass. Around Murchison, the availability of grass varies seasonally: during the dry season grass becomes relatively scarce, in part because of deliberate burning. At this time elephants supplement their diet with the leaves, stems and bark of bushes and trees (Fig. 3.13). This causes the destruction of the trees and bushes and into such areas grass rapidly spreads; it is possible then for fire to spread in and destroy more trees. In such circumstances woodland, even forest, can be replaced by savanna. With the regular alternation of wet and dry seasons grassland becomes more and more dominant relative to woodland. Hence because they tend to destroy the woody part of their habitat elephants in the long run create more grass for themselves, which of course leads to more elephants. The situation here described has not, in my view, been investigated with sufficient insight but there is evidence that under protection elephants are limited in numbers by the availability of grass in the dry season. It seems certain that such limitation occurs density-dependently.

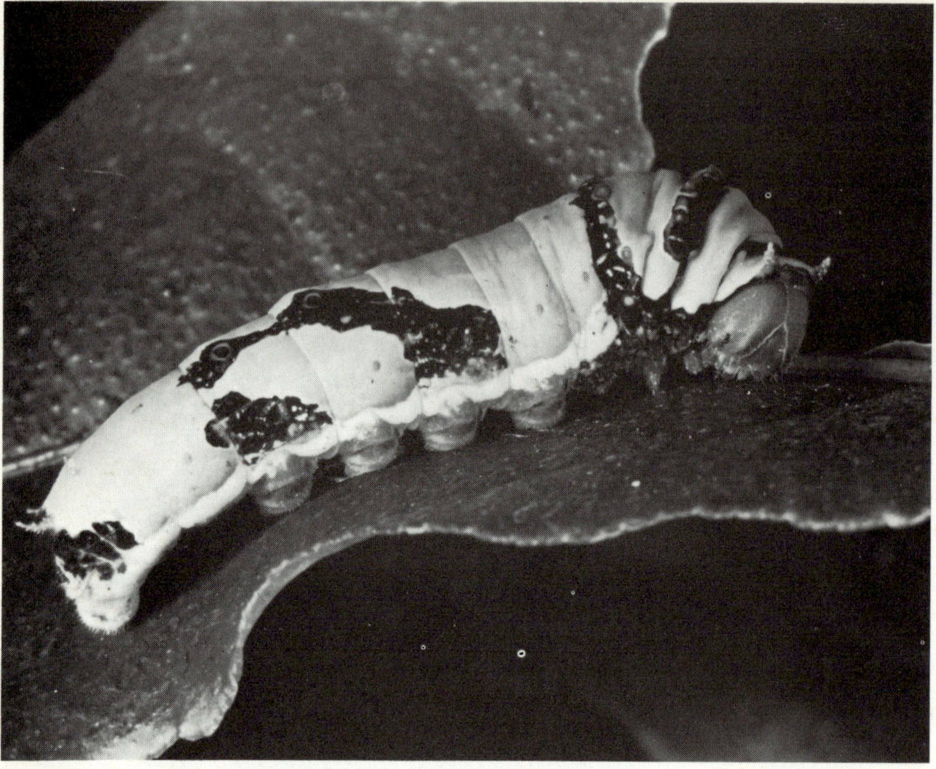

Fig. 3.12 A fully-grown larva of the swallowtail butterfly, *Papilio demodocus*, feeding on an old and tough citrus leaf.

The open-billed stork, *Anastomus lamelligerus*, feeds mainly on the large amphibious snail, *Pila ovata*, but is also reported as eating a variety of other molluscs including *Limicolaria martensiana* (Kahl, 1971). *Pila ovata* becomes particularly abundant in wallows created by the hippopotamus especially if, as is often the case, a floating plant, the water cabbage, *Pistia stratiotes*, covers the surface of the water. The size of a wallow depends on its age and on the number of hippopotamuses using it. Big wallows can be completely covered by *Pistia*, which in turn supports a large population of *Pila ovata*. The storks become concentrated at wallows and feed on the snails, often using the back of a wallowing hippopotamus as a perch from which they can catch the snails. The number of storks at any particular wallow seems to depend on the number of snails, which in turn may depend on the amount of water cabbage, which itself depends on the hippopotamus creating the wallow in the first place. Hippopotamuses, like elephants, have increased under the protection offered them in national parks and their numbers have risen to an extent that they have reduced the carrying capacity of the savanna by overgrazing; they have, in some places, caused severe soil erosion. It is possible, though by no means certain, that the storks are density-dependently food-limited while the snails in the wallows are limited not by *Pistia*, which is very abundant, but by density-dependent predation by the storks. This interpretation is based on qualitative observations and the whole situation is in need of a detailed quantitative assessment, which would not be difficult because each of the species involved is conspicuous and easy to count.

Tsetse flies, *Glossina* spp, feed as adults on the blood of large mammals, especially ungulates. They are never especially common even in areas where there are large concentrations of

Fig. 3.13 An elephant feeding on a bush in the savanna. (Photograph by C. H. F. Rowell.)

mammals. Various people have suggested that the population sizes of tsetse flies are controlled by factors such as climate and some have suggested that the control acts density-independently. The argument is that large mammals supply food far in excess of that required by the number of flies actually found. However, there are two reasons for thinking that tsetse are in fact food-limited in a density-dependent manner. First, the flies seem highly selective in the species of mammals from which they take blood. Table 3.7 shows the relative abundance of large mammals in an area of East Africa and the nature of the gut contents of flies collected in the same area. In this area the most abundant mammal, the impala, contributed only 1 per cent of the food of the flies, while the warthog, which was relatively scarce, contributed 77 per cent. Only one buffalo

Table 3.7 *Relative abundance of large mammals and the food of* Glossina swynnertoni. *(From Glasgow, 1963.)*

Species of mammal	*Frequency of mammals (per cent)*	*Presence of mammal blood in gut of flies (per cent)*
Warthog, *Phacochoerus aethiopicus*	2·68	77
Rhinoceros, *Diceros bicornis*	0·21	2
Buffalo, *Syncerus caffer*	0·02	14
Dikdik, *Rhynchotragus* sp.	7·89	—
Grant's gazelle, *Gazella granti*	2·61	—
Giraffe, *Giraffa camelopardalis*	6·63	—
Hartebeest, *Alcelaphus* sp.	3·51	—
Impala, *Aepyceros melampus*	69·91	1
Lesser kudu, *Strepsiceros imberbis*	0·68	—
Waterbuck, *Kobus* sp.	4·38	—
Other large herbivores	0·46	—
Carnivores	1·01	—
Unidentified Bovidae	—	6
	99·99	100

Note: The percentage frequency of mammals is based upon a mean of 180 mammals seen daily during the scanning of about two square miles (about five square kilometres). 139 fly meals were examined.

was found in the area yet buffalo blood was found in 14 per cent of the flies examined. It is evident from Table 3.7 that not all of the available mammals are utilized by the flies but why this should be so is not entirely clear since the blood of the different species is nutritionally similar. The preferred food of the flies is relatively scarce. Secondly, there is much evidence that as the density of biting flies (including tsetse) rises, mammals scratch and shake themselves more frequently. This may result in the flies being unable to feed properly and hence their numbers may be limited by decreased availability of food as the result of the behaviour of the mammals, which, of course, is stimulated by the rising density of the flies. Both the analysis of fly gut contents and observations on mammal behaviour illustrate the importance of food-availability rather than food abundance; it appears that superficial impressions of a surplus of food may be misleading and that abundant food is not necessarily always available.

The level of oxygen in some tropical swamps is low. This is the result of the abundance of plant material and a corresponding increase in oxygen-consuming micro-organisms. Oxygen-depletion is accelerated by high temperatures, limited movement of the water, and relatively little photosynthesis because of shading from light. The density of oxygen-consuming micro-organisms determines the level of oxygen and they are therefore competing for oxygen among themselves. Whether this results in competition for oxygen in other (metazoan) animals of the swamps is not known, but the possibility exists (Beadle, 1961).

The above examples are suggestive of the widespread importance of density-dependent events and competition in the regulation of animal numbers. A great deal of the evidence is indirect and circumstantial. As already mentioned, density-independent events, such as grass fires and rising levels of water, also act on populations.

1. Lilac-breasted roller, *Coracias caudata*, on a road in East Africa. This and other species of rollers are among the many birds that are attracted to savanna fires where they feed on insects disturbed by the burning.

2. A leopard tortoise in the East African savanna showing how its colour and pattern match the surroundings.

3. Young wildebeest trampled to death during migration in the Serengeti National Park, Tanzania.

4. Mimicry in West African butterflies. The models are all members of the Danaidae and Acraeidae. The sex is given only in cases where there is a difference in colour and pattern. The names of polymorphic forms of mimics, some of them limited to one sex only, are also given. The resemblance between models and mimics is even more striking in living butterflies flying together in the forest. Plates 4 and 5 depict an entire mimetic assemblage in a small area of forest in Sierra Leone. Other assemblages based on different colours and patterns occur in the same forest. (From Owen, 1974b.)

Left

Bematistes vestalis (Acraeidae)
Pseudacraea eurytus form *striata* (Nymphalidae)
Bematistes umbra (Acraeidae)
Bematistes alcinoe male (Acraeidae)
Pseudacraea eurytus form *youbdonis* (Nymphalidae)

Middle

Bematistes macaria female (Acraeidae)
Bematistes alcinoe female (Acraeidae)
Pseudacraea eurytus female form *simulator* (Nymphalidae)
Bematistes macaria male (Acraeidae)
Pseudacraea eurytus male form *eurytus* (Nymphalidae)

Right

Amauris egialea (Danaidae)
Amauris hecate (Danaidae)
Amauris tartarea (Danaidae)
Hypolimnas dubius form *dubius* (Nymphalidae)
Hypolimnas dinarcha (Nymphalidae)

5. Mimicry in West African butterflies (continued from Plate 4)

Left

Amauris niavius (Danaidae)
Papilio dardanus female form *hippocoon* (Papilionidae)
Hypolimnas dubius form *anthedon* (Nymphalidae)
Pitthea famula (Geometridae)
Pseudaletis leonis (Lycaenidae)

Middle

Bematistes epaea female (Acraeidae)
Acraea jodutta female (Acraeidae)
Pseudacraea eurytus female form *eurytus* (Nymphalidae)
Elymnias bammakoo form *bammakoo* (Satyridae)
Papilio cynorta female (Papilionidae)

Right

Bematistes epaea male (Acraeidae)
Pseudacraea eurytus male form *eurytus* (Nymphalidae)
Pseudacraea eurytus female form *epigea* (Nymphalidae)
Elymnias bammakoo form *lise* (Satyridae)
Mimacraea neurata (Lycaenidae)

6–9. In some East African national parks the hippopotamus has become so common that it is necessary to reduce the population by cropping. Students and staff from Makerere University in Uganda have exploited the cropping and by detailed examination of the dead animals have discovered a great deal about the ecology of the species.

6. Examining the teeth in order to estimate the age of an animal.

7. Students dissecting the stomach region.

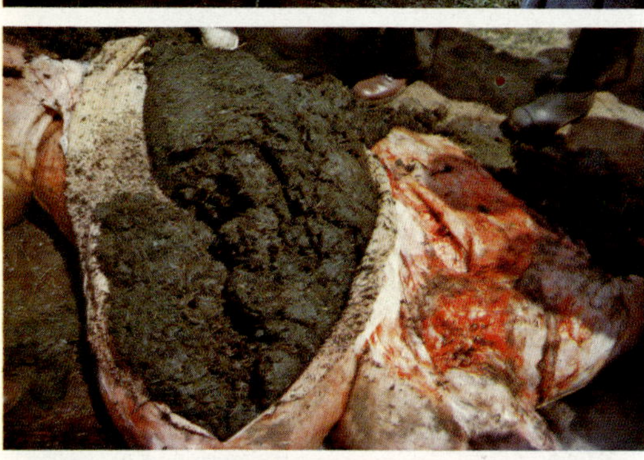

8. The stomach contents are almost entirely grass and look like lawn mowings.

9. Searching the eye for the parasite *Oculotrema hippopotami*.

10. New leaf growth on a recently burnt *Lophira* tree. Certain species of trees and many grasses produce new growth soon after a fire. Sierra Leone.

11. The Virunga Volcanoes as seen from Lake Mutanda, Uganda. A beautiful and varied region with an overwhelming variety of species of plants and animals.

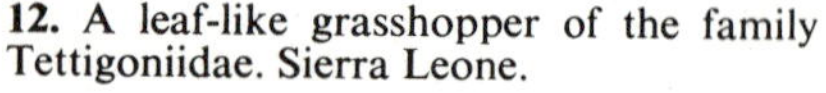

12. A leaf-like grasshopper of the family Tettigoniidae. Sierra Leone.

13. A trap set to catch clawed toads, *Xenopus*, at Lake Mutanda, Uganda. The trap is baited with termites and the toads enter it through a funnel. In this part of Uganda *Xenopus* is considered a delicacy; they are dried in the sun and eaten whole.

14. A large snail of the family Achatinidae. Snails of this family occur throughout tropical Africa. There are many species and locally distinct forms and there is much confusion over how many species there are in Africa.

15. Dry savanna in Karamoja, Uganda. A patch of *Aloe* is growing in the shelter of a *Euphorbia* tree.

16. Savanna in Sierra Leone dominated by *Lophira* trees. The wet season.

17. The same area in the dry season showing partial burning of the trees and grass.

18. Praying mantis, *Pseudocrea bothrea*. Throughout Africa mantes are among the more conspicuous predators of insects. This one feeds on butterflies. Uganda.

19. Tree frogs of the family Rhacophoridae in dry savanna in northern Sierra Leone. Africa is extremely rich in species of tree frogs but most species occur in rain forest.

20. The butterfly, *Cyrestris camillus*, one of the many species that has exploited cultivated areas in Africa.

21. Flowers of *Marica caerulea* in a garden in Uganda. The flowers last for one day only. They are produced at intervals of about three days in all months of the year and this could be an example of an endogenous rhythm, as described on Page 80.

The balance between birth and death rates

All animals are capable of exponential increase in numbers and yet, apart from relatively minor fluctuations, most populations are maintained at stable equilibrium. This means that births and deaths are balanced and that in most populations the death rate, particularly among young individuals, is extremely high. There is of course much variation in the birth rate between species – an elephant produces one young at a time; the land snail, *Limicolaria martensiana*, lays a clutch of about 15 eggs; the butterfly, *Acraea encedon*, produces 200 or 300 eggs – and it must be assumed that there is corresponding variation in the death rate in these three species.

Is the birth rate adjusted to the 'expected' death rate? In view of the stability of most populations there is undoubtedly a mathematical relationship between the two, but is there a biological one and if so how is it achieved? Alternatively, perhaps birth and death rates are independently determined by the availability of resources; that is to say, the death rate balances the birth rate but the birth rate is not specifically adapted to the death rate. These two possibilities have an identical outcome – a balanced population – but there are ecological and evolutionary differences between them. Here I shall discuss the problem by reference to birds.

Most tropical birds, especially passerines, lay smaller clutches than similar species in temperate regions. Clutches of two eggs are common among tropical species while their high latitude counterparts lay five or six eggs. Indeed there is a gradient of increasing clutch size from the equator to the polar regions. Does this mean that the lower clutch sizes of tropical birds have evolved because the death rate of birds is lower in the tropics? Or does it mean that in the tropics parent birds experience greater difficulty in providing food for their families, and hence the evolution of small clutches? In the majority of birds – there are exceptions – the parents look after and feed the young in the nest. At high latitudes birds breed in the spring and early summer when the days are long. It has been suggested that the increased clutch sizes characteristic of high latitude species represents an adjustment to the number of young for which the parents can, on average, provide food; in other words, long days and a superabundance of food for a short period of the year make large families possible. But in the tropics daylength is relatively constant and food is rarely excessively abundant; hence the evolution of small clutches. Moreover, compared to high latitude species, tropical birds are faced with an enormous diversity of potential food which probably makes food-gathering difficult. At high latitudes an insectivorous bird is able to concentrate on rather a small number of species of insects while a tropical bird is faced with a large number of species, few of them common. Birds may be described as 'conservative opportunists'; they learn by experience and are able to build up search images of what they are looking for. It therefore follows that it is easier for a bird to obtain food in an area where there are a few common species and much more difficult where the potential food is more varied.

Studies of variations in the reproductive rate in birds support the view that birth rates are adjusted to the average availability of resources at the time of breeding and do not compensate for death rates. The same may be true of other animals, even those without parental care; such animals probably produce the maximum number of eggs or young that is physically or physiologically possible. Birth rates and death rates are probably independently adjusted to the available resources, in particular to the availability of food. Numbers of animals are not, in general, controlled by variations in birth rates but by variations in death rates, usually acting density-dependently.

The effect of overcrowding on body size

In most animals a rise in population density and a shortage of resources usually lead to an increase in the death rate, but in a few species may result in a decrease in average body size. If, for example, the larva of a butterfly depletes its food-plant it often dies but alternatively it may pupate before it has reached full size and eventually produce a smaller than average butterfly.

Growth inhibition as the result of overcrowding, even in the presence of abundant food, is well known in laboratory populations of aquatic snails, tadpoles, and other aquatic animals. These animals evidently release a growth inhibiting substance that has a marked effect on other members of the population. Larger individuals release relatively more of the substance and suppress growth in smaller individuals. *Biomphalaria sudanica*, an African aquatic snail, releases a substance that appears to inhibit growth in other snails in the same population (Berrie and Visser, 1963). In this species maturation occurs only when the snails reach a certain size and so breeding is delayed at high population densities.

The stimulus to release a growth inhibitor is density-dependent but a similar effect could occur without the release of a specific substance. Thus populations of fish at high density contain many individuals that are smaller than the average for their age-group. This arises partly from food shortage and partly perhaps from a shortage of space. But fish reproduction is often independent of size and no matter how crowded the population the fish continue to breed.

Competition and ecological segregation

If competition for resources is important in determining the size and density of populations it might be expected that mechanisms would evolve that would tend to reduce, to some extent, the effects of competition. Ecologists since Darwin have repeatedly drawn attention to the fact that similar species of animals usually differ in their ecological requirements. Thus in a group of related species it is usually possible to find differences in morphology that reflect differences in ecology, especially in the kind of food eaten. Such ecological segregation between species is thought to be an evolutionary consequence of interspecific competition for limited resources: if species have different requirements there is less likelihood of competition. Many examples of ecological segregation are known from tropical African animals. Thus, Moreau (1948) examined ecological differences in 172 species of birds living in an area of 3 000 miles^2 (7 770 km^2) in Tanzania. He found that 94 per cent of the related species (species in the same genus) differ in diet or in habitat, or in both. His analysis excluded the weaver birds, the species of which show greater similarity than other birds. Most weavers feed on temporarily superabundant grass seeds and in the dry season when the seeds become scarce the weavers move to other areas of temporary superabundance.

There are in Africa two species of *Schistosoma* that commonly cause schistosomiasis in man; additional species are known from other mammals. One species that infects man, *S. haematobium*, depends on aquatic snails of the family Bulinidae for its intermediate hosts. These snails are most abundant in pools of water in the drier areas of Africa and they are able to resist and withstand temporary drying up of their habitat. Hence infection with *S. haematobium* is especially associated with the drier, savanna areas of Africa. *Schistosoma mansoni*, on the other hand, depends upon snails of the family Planorbidae for its intermediate hosts. These snails are more abundant in the wetter parts of Africa. Thus the two species of *Schistosoma* are ecologically segregated by habitat and by the intermediate hosts they use. But there are additional differences between the two species: *S. haematobium* infects the blood vessels of the urinary system, while *S. mansoni* infects the blood system of the intestine. The two parasites thus derive their food from different parts of the human body. A person can be infected with both species though generally this is not so because the parasites and their intermediate hosts occur in geographically and ecologically separate areas.

There are 147 species of the butterfly genus *Acraea* in Africa. The genus includes forest and savanna species and an extraordinary number of species have become adapted to cultivated land and gardens; indeed they are among the most familiar butterflies in Africa. Gardens within the forest region of West Africa support not only forest species of *Acraea* but also savanna species that have colonized and exploited the changed environment.

Twenty-one species of *Acraea* are known from the immediate vicinity of Freetown in Sierra

Leone. Although all the species are similar in appearance each one is distinctive in its ecological requirements. The larval food-plants of 14 species are known and, with one exception, each utilizes separate plants, in many cases plants belonging to quite different families. The adult butterflies are nectar-feeders, and although many of the species fly as adults throughout the year each tends to reach a seasonal peak of abundance for only a restricted period of the year. This seasonal pattern may be associated with seasonal changes in the availability of larval food-plants, but one effect is that the timing of the peak abundance of each species is staggered. The *Acraea* butterflies are further segregated by habitat (some require shade, others open country) and by the height at which they exploit nectar sources. It is difficult to explain these and other ecological differences between members of a group of closely similar species other than in terms of adaptations that minimize interspecific competition for resources.

Conclusion

A feature of population regulation in tropical Africa and in other tropical areas of the world is that the factors controlling a given population are, in many cases, highly complex. Many more species occur within a single area than in temperate regions; species tend to have more varied diets and are preyed upon by a larger variety of predators than in temperate regions. Simple predator-prey situations seem rare. Few cases of population regulation have thus far been analysed quantitatively and most of what has been said in this chapter is descriptive and speculative; there is scope for detailed investigation and a start could be made with relatively simple examples, such as the hippo-snail-stork association already described.

Tropical populations seem more stable than temperate populations. Cyclic fluctuations in numbers are not known to occur and it is easy to form the impression of an extremely complex environment that imparts considerable long-term stability, a point that will be raised again in the next chapter.

Chapter 4

Communities and ecosystems

Species of plants with similar habitat requirements tend to occur together. The assemblage of plants in a locality and the animals associated with them is called a community. An ecosystem is a community together with the non-living part of the environment.

The kind of community that develops in a particular locality depends on the physical features of the environment. The type of soil, the amount and seasonal distribution of rainfall, temperature, and variations in daylength are examples of factors that determine what sort of community occurs. As would be expected, communities tend to repeat themselves in different parts of the world, although it is not always exactly the same assemblage of species of plants and animals that is involved in each locality. Sometimes the presence of a particular mineral in the soil severely restricts colonization by plants; at old mine dumps in Zambia where the soil contains more than 1 per cent of copper only two species of plant, a labiate, *Becium homblei*, and a grass, *Chloridion cameronii*, are able to establish themselves (Reilly, 1967). In this area no other plants can tolerate such a high concentration of copper and it must be assumed that in communities like this there are very few species of animals.

An experienced ecologist can tell from the presence of certain species of plants what minerals are in the soil, whether the soil is acid or alkaline, and how much rain falls in the area. Farmers and cultivators can often judge from the natural vegetation what crops are worth growing on a piece of land; indeed most people can make at least some predictions about the agricultural potential of a piece of land by simply looking at it – but of course the reliability of such intuition varies according to previous experience.

The assemblage of plants and animals in a community tends to remain at stable equilibrium year after year; that is to say, common species remain common and there is no appreciable change in the frequency of rare species. But if a community is violently disrupted by a catastrophic event its essential qualities disappear and it may take many years for them to be recovered. Well-established and stable communities are usually referred to as climax communities. The word climax is used to imply that the community has stabilized itself and has developed an assemblage of plants and animals that represents the maximum possible (in terms of diversity and abundance) under the particular conditions of climate and soil of the locality in question. Tropical forest undisturbed by human activities is the most complicated climax community in the world, if viewed in terms of the diversity of species and the interactions between them.

If a community is disrupted there begins almost immediately a series of sequential changes which, given time, lead to the restoration of the original community or, if the disruption has altered fundamentally the characteristics of the area, a new kind of community is developed. At Lamto in the Ivory Coast a severe grass fire immediately reduces arthropod numbers by 36 per cent and reduces their biomass by as much as 68 per cent (Gillon, 1970); 2 months later arthropod numbers and biomass are restored and it is assumed, in this instance, that the grassland is well-adapted to periodic fires and that the community is in effect re-established every year (Gillon and Pernes, 1968).

The word succession is used to describe the changes that occur after a community has been suddenly altered. At Lamto, where there are grass fires every year, the same pattern of succession is repeated year after year. This is not always the case, for a totally new kind of succession can occur particularly where man has initiated a change. In 1964 the dam across the Volta River in Ghana was closed and the Volta Lake, now one of the biggest man-made lakes in the world, began to form. Thus part of a river was transformed into a lake which eventually covered an area of about 7 000 km^2. Many of the plants and animals present in the Volta Lake are derived from those that occur in the Volta River. The lake is still undergoing succession and no doubt will eventually reach a stable climax condition. Changes have been reported in species-composition and species-abundance of the animals (especially fish and aquatic insects) in the lake but no one knows for certain exactly what kind of community will finally establish itself. In this instance, in contrast to the seasonal burning of grassland, succession will produce a community quite different from what used to exist in the area.

The intricacies of food webs

A community requires the presence of green plants which by the process of photosynthesis convert radiant energy from the sun into chemical energy. Plants provide food for animals and in all communities there is an intricate series of feeding relationships that can be summarized by writing,

green plant ⟶ herbivore ⟶ predator.

However, this is a simplification as there are usually more predators (though rarely more than four) that extend the links on the right of the chain. Moreover, all species of plants are utilized as food by more than one species of herbivore and these in turn are eaten by a variety of species of predators, which themselves are eaten by predators. Because of these intricate feeding relationships it is customary to speak in terms of food webs. Indeed it is likely that all the species in a community are connected to one another in terms of the acquisition of food. It is, therefore, impossible to construct a complete food web for a community like the Volta Lake or for an area of grassy savanna, although it is possible to work out the more common feeding relationships that occur.

To build up a complete picture of a food web in a community it would be necessary to know, for each species, not only what items are eaten but also how often they are eaten. It is known, for instance, that the larvae of many butterflies feed on a variety of related species of plants, but it is not known, in most cases how often they utilize each species. Similarly, insectivorous birds are known to eat a variety of species of insects; each species of bird exercises a distinct preference, but although there is abundant qualitative evidence as to what birds eat only rarely is there quantitative information. There are two ways by which information about food webs may be obtained: examination of gut contents and direct field observation of feeding behaviour. Both methods have their disadvantages: the contents of the gut are likely to be digested at different rates and hence some items tend to be over- and others under-represented; field observations necessarily restrict the precision with which prey items are identified – it is not difficult to see that certain birds eat dragonflies, but much more difficult to identify the dragonflies to species and to score the frequency at which various kinds are taken.

In most parts of Africa at least one species of shrike occurs in gardens. These conspicuous and familiar birds are predators and they take an enormous variety of prey, chiefly large insects, but also small vertebrates. Many of the insects taken by shrikes are themselves predators while others feed directly on plants. Shrikes often catch and eat praying mantids which feed on other insects of a great variety of species. Mantids are occasionally attacked and eaten by geckos, but a gecko is often about the same size as one of the larger species of mantid and now and again the potential predator ends up as the prey. Any attempt to construct a food web centred around a shrike feeding in the garden would necessarily be so over-simplified as to be almost meaningless:

all that can be said is that shrikes are opportunist and take anything of the appropriate size they happen to see. Many of the species (including mantids) that constitute the prey of shrikes are themselves opportunists and it is really only the plant-feeders that are relatively restricted and conservative in their choice of diet, though even with these there are species that are able to utilize a wide range of plants.

The driver ants, *Dorylus*, with their foraging swarms, are group predators and their behaviour includes both group raiding and group retrieval of prey. These insects, exploiting almost anything living – chiefly insects, spiders, and earthworms – are capable of overpowering quite large vertebrates (including snakes) and even feed on nuts and seeds. Lists of prey taken by driver ants (Gotwald, 1974) reflect primarily the diversity of animals that occurs in a given community. Driver ants are unspecialized in what they eat but their method of obtaining prey – sending out densely-packed foraging columns and overpowering anything they encounter – is highly specialized. Perhaps more than any other group of animals they illustrate one of the axioms of community ecology: all events are inter-related.

Production in ecosystems

An ecosystem, like a community, can be large or small – it is a matter of convenience where the line is drawn. At one extreme it is justifiable to speak of the whole of the living world as one gigantic ecosystem but it is more useful and manageable to think in terms of the forest ecosystem, the savanna ecosystem, and so on; indeed it is reasonable to speak of an ecosystem centred around a single living plant. The essential feature of an ecosystem is that it provides a means of converting the sun's energy into stored chemical energy. This is normally achieved by green plants through the process of photosynthesis but there are a few other organisms, including red and brown algae and certain bacteria, that are also capable of converting the sun's energy into chemical energy. Unless the means of conversion are present there is, by definition, no ecosystem: an area of arid desert or a slab of lateritic rock devoid of plants do not provide the essential requirements of an ecosystem.

The productivity of an ecosystem is measured by the rate at which chemical energy is produced during photosynthesis. The total amount formed in a given time is called the gross primary productivity and the amount left over after utilization by the growing plants is the net primary productivity. It is the net primary productivity that is a source of food utilized by animals and by some parasitic plants.

The biomass of plant material present at a particular time is sometimes called the standing crop and is usually expressed in terms of dry weight per unit area. The greater the biomass of plant material the greater the biomass of dependent animals. Everything produced by the plants is eventually consumed, there is no long-term accumulation of dead material, and it is the details of exactly how plant production disappears that is of special interest to ecologists.

Sunlight, carbon dioxide, water, and salts in solution are all that is required for plant production to occur, but the rate of productivity varies with the availability of these necessities – except perhaps carbon dioxide, which is usually present in sufficient quantity. In the tropics there is not the seasonal variation in available light characteristic of high latitude areas; nevertheless even in the most humid equatorial environment some seasons of the year are better for plant production than others. Land plants require considerable quantities of water for photosynthesis (and for other life processes) and in many tropical areas it is a shortage of water that inhibits and restricts plant production. In most of Africa there is at least one dry season a year during which plant production is much reduced. Plant production, like all biological processes, is affected by temperature but in the tropics this cannot be considered as a major limiting factor.

Net primary productivity and biomass have been estimated for all the major world ecosystems. In Table 4.1 the estimates for tropical and temperate forest and grassland are

Table 4.1 *Estimated productivity and biomass for forest and grassland in tropical and temperate areas. (From Owen, 1974a.)*

	Average net primary productivity ($g/m^2/year$)	*Average biomass (kg/m^2)*
Tropical forest	2 000	45
Tropical savanna	700	4
Temperate forest	1 300	30
Temperate grassland	500	1·5

compared. These figures are world estimates and as such are subject to considerable regional variation. The net primary productivity of tropical forest is the highest in the world and is rivalled only by swamp, marshland, and estuaries. It is about four times as high as the world average for agricultural land and, as shown in Table 4.1, is nearly three times as high as savanna. Net productivity for temperate forest is not substantially smaller than for tropical forest and that for temperate grassland is only slightly smaller than tropical savanna. In terms of biomass, tropical forest supports the greatest in the world; this is because the trees are large and there is much woody material. The biomasses of tropical savanna and temperate grassland are much less, mainly because of the scarcity or lack of woody vegetation and the predominance of the herbaceous layer, chiefly grass.

Table 4.2 gives some estimates of net primary productivity for the herbaceous vegetation (all

Table 4.2 *Minimum estimates of net primary productivity based on above-ground standing crop data for herbaceous (non-woody) vegetation in African savanna ecosystems. From Bourlière and Hadley (1970).*

Locality	*Annual rainfall, in mm*	*Growing season (days)*	*Net primary productivity ($g/m^2/year$)*
Olokemeji, Nigeria	1 168	270	680
Shika, Nigeria	1 118	200	340
Richard-Toll, Senegal	200–400	60	40
Lidney, Chad	*c.* 320	40	37
Mohi, Chad	*c.* 320	40	125
Şerengeti, Tanzania	*c.* 700	?	520
Kivu, Zaïre	860	?	1 750

plants except trees and shrubs) for various savanna ecosystems in tropical Africa. Each area is characterized by the dominance of one or more species of grass and the main difference between them is in the annual rainfall and the length of the growing season, which, of course, is determined by both the amount and seasonal distribution of rain. As shown, one of the estimates is higher and the remainder lower than the world estimate for tropical savanna given in Table 4.1, but all the figures in Table 4.2 exclude woody vegetation, which, at least in some places, may account for a considerable amount of the net primary productivity. The highest net primary productivity known from African savanna occurs in the *Imperata*-dominated grassland at Kivu, Zaïre, while the lowest estimates are from the arid Sahelian zone in Chad. The correlation between productivity and annual rainfall is impressive and it must be assumed that in the African savanna the chief limiting factor to productivity is the availability of water. During the recent Sahel drought it is likely that the productivity of arid savanna areas was close to zero. The variations shown in Table 4.2 undoubtedly determine and restrict the abundance and diversity of animal life. In East Africa the biomass of ungulates and their predators has been determined in several areas, but for a more complete picture it is necessary once again to return to the excellent work at Lamto in the Ivory Coast, many aspects of which are summarized by Bourlière and Hadley (1970) in their review of the ecology of savanna ecosystems.

In this area, described as forest-savanna mosaic, seasonal rainfall is all-important. At the height of the dry season in January there is effectively no living vegetation above ground and almost the entire biomass of dead vegetation is suddenly removed by fire. Plant growth recommences as the rains begin in February and March and peak biomass is achieved at the end of the wet season in September and October. The herbaceous plants begin to die (but remain standing) in November and by the end of December almost everything above ground is dead. Arthropods reach their peak biomass at the end of the wet and beginning of the dry season. About two-thirds of their biomass consists of species that feed directly on plants; this fraction is maintained, with some irregular fluctuations, throughout the year. But although most of the biomass consists of plant-feeders, most of the species are predators and parasites. In other groups, particularly snakes and rodents, peak biomasses also occur at the end of the wet and beginning of the dry season, but the highest biomass of amphibians occurs earlier in the wet season in May and June. The biomass of birds varies less, on a seasonal basis, than that of other groups of animals but tends to be higher in the wetter months of the year. At Lamto, then, seasonal variation in plant production and the relative quantities of living and dead vegetation available at different times of the year, have profound effects on the animal community, and the entire ecosystem seems to be adjusted to seasonal rainfall and, of course, to grass fires in January.

Energy entering an ecosystem by the process of photosynthesis is eventually lost in the form of heat. There may be a certain amount of temporary recycling of energy but the end result is a flow of energy in and out of the system. In order to live organisms have to 'work' and this always results in a loss of heat. Every time an animal eats plant material some of the energy is converted into animal tissue and some is lost as heat or as indigestible waste products. Exactly the same is true when an animal eats another animal. Plants and animals are remarkably inefficient in their use of energy. It is possible to measure the energy content of sunlight and the energy content of plants can be determined by burning the material completely and measuring the heat (in joules) that is given off. The ratio of the energy in plant material to that in incident sunlight provides an estimate of the efficiency of plants and turns out to be very low, not more than 1 per cent for most plants but with considerable variation. Similar estimates may be made by measuring the energy content of animals and the plants they eat. Growing animals are more efficient at making use of the energy of their food than are fully-grown adults – indeed unless an animal is growing it consumes an amount of food that provides only enough energy for day-to-day maintenance. For many adult insects all that is needed is a certain amount of refuelling, but in order to obtain food (which is often the sugary nectar or pollen produced by flowers) they have to expend energy in flying to its source.

It is tempting to speculate that the various stages of succession in a community would lead to increasing productivity and this is often, but not always, the case. It is perhaps better to view a climax community in terms of equilibrium between production and consumption and not necessarily in terms of productivity. In the next section the relationship between what is produced and how it is consumed is explored more fully.

Trophic levels and trophic models

Every time an organism eats another organism there is both a transfer and a loss of energy. With this information it is easy to predict that the biomass of plants will be substantially greater than that of the animals that eat them – a prediction that no one would seriously disagree over, except perhaps for certain aquatic environments where the generation time of the producers is short and their biomass is never particularly high. It follows, too, that the biomass of predators should be smaller than that of their prey; indeed there ought to be a quantitative decrease in biomass up a food chain so that the top predators in an ecosystem are relatively rare as individuals and low in biomass. Detailed study of selected communities, in Africa and elsewhere, has confirmed these predictions.

The various steps in a food chain or web are called trophic levels, a term that has about the same meaning as feeding levels but is more inclusive because it takes into account the acquisition of energy by plants, a process that is difficult to envisage as 'feeding'. The trophic levels in an ecosystem may be listed as follows:

(a) *Producers*, normally photosynthetic plants which convert the sun's energy into stored chemical energy.
(b) *Primary consumers*, usually plant-feeding animals (herbivores) and sometimes called secondary producers.
(c) *Secondary consumers*, carnivorous animals, commonly referred to as predators and parasites, that feed on primary consumers. A few insectivorous plants (like sundew) are secondary consumers as well as producers.
(d) *Tertiary consumers*, animals that feed on secondary consumers, also called predators and parasites.
(e) *Higher-order consumers*, all other predators and parasites that feed chiefly on tertiary consumers, although there are animals, relatively few in number and low in biomass, that occupy even higher trophic levels.

These distinctions refer to feeding levels and not necessarily to the characteristics of individuals, populations, or species. The producers can, however, be defined with confidence as very few plants obtain energy other than by photosynthesis. The majority of species that make up the primary consumers always feed at this level – it is relatively unusual for an animal to take both plant and animal food – but there are exceptions, like driver ants, which besides being secondary and tertiary consumers, are also primary consumers when they eat seeds and nuts. Most birds are secondary, tertiary, or higher order consumers, and an individual often switches between trophic levels as it takes successive items of food.

It is useful to distinguish between consumers feeding on living organisms and those feeding on dead organisms. Dead producers are no longer capable of photosynthesis and the consumption of dead plant material has quite a different effect on an ecosystem than the consumption of living plants. A variable, but generally small, fraction of plant material is utilized while it is alive, the remainder goes to primary decomposers which are in the same trophic level as primary consumers but differ in that they make use of dead plants.

Trophic levels can be represented by constructing a model for the transfer of energy in an ecosystem. There are several possibilities: the one I favour is where a distinction is made between whether an organism is eaten while it is alive or after it has died. In this model (Fig. 4.1) the trophic pathways of consumers and decomposers run parallel. Thus primary consumers are eaten by secondary consumers (Fig. 4.2) or, if they die, by secondary decomposers which can also feed on dead primary decomposers. In Fig. 4.1 '*n*th order' refers to tertiary and higher order consumers and decomposers and is written like this to save space and repetition. At each level there is heat loss and all the energy entering the system eventually leaves it. The model does not allow for organisms that as they feed switch between levels or those that are partly decomposers and partly consumers at the same level, but the role of an individual or a species in a community is not in this context as important as its trophic position at the time it acquires food.

How many trophic levels can be expected in an ecosystem? There are normally at least three, often four, but beyond this number the biomass of organisms involved is so small that they play an unimportant role so far as energy transfer is concerned. This in effect means that birds like eagles, although large and conspicuous, are relatively unimportant in most ecosystems.

The bulk of the primary consumers in a forest ecosystem are insects. And in forests in particular insects have little impact on living vegetation – widespread defoliation is rare – but once the vegetation dies it quickly disappears through the actions of the decomposers. In West

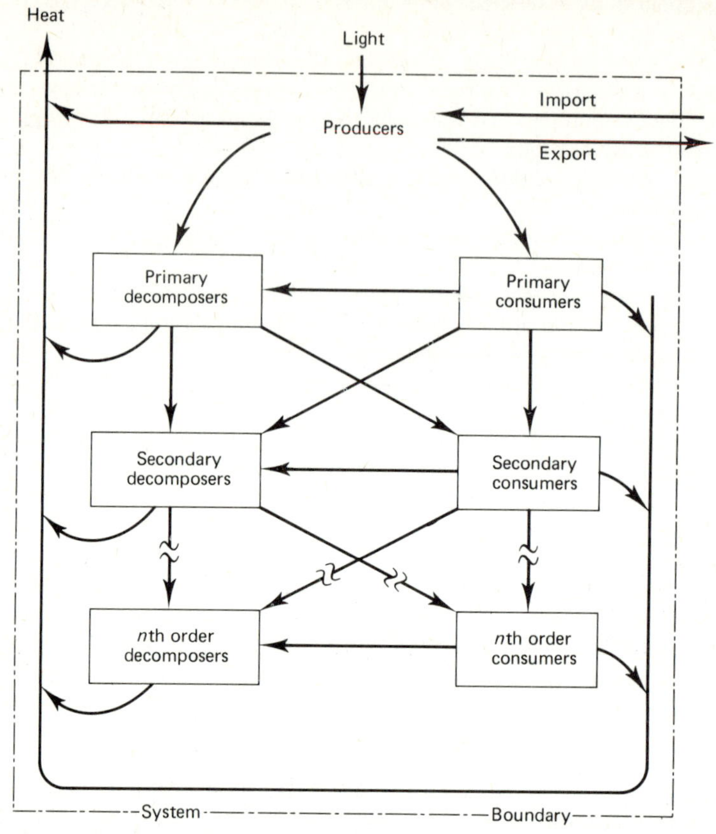

Fig. 4.1 A model for the transfer of energy in an ecosystem. (From Owen, 1974a.)

and Central African rain forests it is likely that less than 5 per cent of the vegetation is eaten while it is alive. The same is true of most savanna areas (for instance at Lamto) although in some areas large ungulates play an important role as primary consumers. In East Africa between 28 and 60 per cent of the primary production of some savanna areas goes to the primary consumers the most obvious of which are the large grass-feeding mammals, although the impact of insects like grasshoppers is largely unexplored in these ecosystems.

These figures of primary consumption contrast markedly with open-water ecosystems. Here the producers are supported by the water itself and are mostly small and planktonic. The level of primary consumption is high, often approaching 100 per cent in the ocean which, of course, means that the primary decomposer level is not well developed; indeed most of the decomposers in the ocean and in large lakes feed on dead primary consumers. The difference between terrestrial and aquatic ecosystems seems to arise from the nature of the difference between air and water. Water physically supports producers and as long as they are small or bouyant they need no special structures to keep their tissues in sunlight. Land plants not only need a substrate for attachment but also need structures to hold their leaves in the light, often in competition with other plants; this probably explains why the majority of land plants are large and woody.

In tropical forests leaves fall throughout the year but especially during the dry season. Each species of tree tends to have a peak leaf-fall at a certain time of the year. The rate of

Fig. 4.2 *(a)* A primary consumer: the larva of the butterfly, *Catopsilia florella*, feeding on the leaves of *Cassia alata*; *(b)* A secondary consumer: a crab spider feeding on a butterfly, *Hypolimnas misippus*. Crab spiders lurk in flowers and seize butterflies that visit the flowers for nectar.

disappearance of fallen leaves is high compared to temperate forests, especially in the wetter months. At one site in Nigeria fallen leaves are initially decomposed by mites and Collembola (Madge, 1965). One species of earthworm, *Hyperiodrilus africanus*, is common at this site but, unlike many temperate earthworms, does not attack fallen leaves directly and feeds mainly on leaves fragmented by other organisms. In forests, savanna, and indeed in virtually all lowland ecosystems in tropical Africa, termites play a dominant role in the decomposition of dead vegetation and it is probably the abundance of termites, together with high temperatures, that results in the rapid disappearance of leaf litter and in a poorly developed humic layer in most tropical soils, particularly in forests. Indeed in forests the bulk of the nutrients of the ecosystem are locked up in living vegetation, in contrast to the situation in temperate forests where there is normally a thick layer of humus. Termites are less abundant at high elevations and disappear altogether high on mountains. On the grassy prairie near the top of the Nimba Mountains in Guinea (which rise to over 1700 m) the main decomposers are earthworms (Lamotte *et al.*, 1962). In this area the primary consumers have hardly any impact on the producers which has led to the belief that the grass is 'under-exploited' by animals; I shall return to this belief later.

Cycles of essential materials

Energy enters ecosystems and is transferred through trophic levels and lost. The intricacies of food webs can be viewed essentially in terms of the acquisition and movement of energy in the form of food. But there is more to ecosystems than the transfer of energy: plants and animals require other materials, including: water; nitrogen from which to form protein; carbon dioxide; and a whole variety of inorganic materials which are mostly needed in small quantity but are nevertheless vital for life processes. These materials circulate again and again through ecosystems and under normal circumstances are never 'lost.'

In Africa the availability of water is perhaps the most important single factor that limits the productivity of land plants. Water is absorbed through the roots and in the dry season there may be little or no water near the surface of the soil and plant growth is inhibited or stops altogether. Most savanna plants are well-adapted to seasonal variations in the availability of water and there is a repeating pattern of growth, flowering, fruiting, and decomposition. The availability of water determines the distribution and abundance of the majority of species of plants and throughout all but the wettest parts of Africa it restricts, at least to some extent, the diversity of species that can occur in an ecosystem. The distribution, abundance, and diversity of animals may also be directly affected by the supply of water but much more often by the availability of plants. Periodic drought resulting from the failure of the rains over several consecutive seasons has disastrous effects on vegetation (and hence on animal life) over vast areas of Africa, especially in the Sahel region where savanna can within a few years be temporarily converted to desert.

Much of the water falling on the land returns directly to the atmosphere through evaporation. In hot savanna regions even quite heavy rainfall is insufficient to sustain the growth of many plants. Indeed rain that in the temperate region would provide enough water for a rich flora would be inadequate in the tropics because of rapid evaporation and return to the atmosphere. Plants take up and lose large quantities of water and relatively little is retained. Animals lose water through evaporation from their body surfaces and in their waste products, and all is returned to the atmosphere. There are, then, two water cycles, one involving direct evaporation from the ground and the other in which the water passes through organisms.

It would be possible to describe the cycles of numerous essential materials but the details of how these occur can be found in most ecology text-books, and, with a few minor exceptions, what happens in the tropics is no different from what happens elsewhere in the world. I shall instead outline the circulation of nitrogen, which, for animals and plants, is an element of profound importance. Few organisms can make direct use of atmospheric nitrogen. It is necessary for it to be converted into a more complex compound before it can be utilized. When dead plants and animals decompose simpler compounds are formed, among them ammonia produced from the decomposition of protein. Ammonia is soluble in water and can be taken up by the roots of plants, and of course returned again when the plants (or the animals that have eaten them) themselves decompose. Some ammonia goes through a different cycle. There are in the soil two groups of bacteria which convert ammonia into nitrites and nitrates. Nitrites cannot be used by the majority of plants but nitrates, taken up in solution, are one of the more important sources of the nitrogen required by plants for the formation of protein. There is another group of bacteria that are able to convert atmospheric nitrogen into ammonia. Some of these bacteria are free-living, others occur in the roots of plants, particularly those of the pea and bean family. The bacteria cause distinct swellings, called nodules, in the roots, and the presence of species of plants with root nodules and their associated nitrogen-fixing bacteria results in local enrichment of the soil. Occasionally, during thunderstorms, the lightning converts atmospheric nitrogen directly into nitrogen compounds which can then be used by plants. There are, then, several ways by which plants acquire the nitrogen they need; animals, of course, obtain their requirements by eating plants and other animals.

Theoretical considerations

As a conceptual framework the idea of the ecosystem is both useful and attractive. The apparently well-organized way in which energy enters and leaves the system, the way plants and animals are arranged into trophic levels, and the way essential materials are recycled, has considerable appeal to ecologists seeking order in the living world. There are, however, some unresolved problems and in this section I shall draw attention to those that seem to me the most interesting.

First, why is it that in terrestrial ecosystems and particularly in tropical forest, the most complex ecosystem of all, the producers are apparently under-exploited by the primary consumers? Initially it would seem that primary consumers, as a trophic level, are not limited in numbers by food; that is to say their population sizes are controlled not by the availability of resources but in some other way. If so, terrestrial ecosystems differ from ocean ecosystems where primary consumers account for almost all plant production. One possibility is that terrestrial primary consumers are limited in numbers by predators and parasites (secondary consumers) and that because of the rate at which they are eaten are never able to build up populations that are near the carrying capacity (defined in terms of available food) of the environment in which they live. This view is favoured by some ecologists but rejected by others. Primary decomposers, on the other hand, account for all dead plant material – there is no accumulation – and it must therefore be assumed that as a trophic level they are food-limited.

Another possibility is that tendencies on the part of primary consumers to exploit more fully the living plant material are countered by the plants themselves. There is in plants a wide array of adaptations that can be interpreted only as defence mechanisms against consumers. These include thorns, tough leaves, downy or hairy stems and leaves, and the presence in tissues of a quite extraordinary variety of toxic compounds which play no part in the normal physiology of plants but which deter animals from feeding on them. Many plants are in effect loaded with natural pesticides and some of them, like pyrethrum, are extracted and used commercially. These adaptations can be understood in terms of the need of land plants to be large and complex with much supporting tissue. Without them primary consumers would presumably never enable plants to develop complex structures and the consumption of living tissue would be much higher than it is – indeed land plants as we know them could not exist. This possibility supports the view that primary consumers, as a trophic level, are food-limited and acknowledges that the bulk of plant material is inedible while it is living. The fact that the larvae of many species of Lepidoptera are able to eat only the new leaves of their food-plants suggests that older leaves are too tough or toxic and do not constitute exploitable food.

The relationship between plants and their consumers is a good example of co-evolution: over millions of years plants have evolved mechanisms that restrict consumption, and yet these same mechanisms are used by some consumers as a means of locating the correct species of food-plant. Thus for most consumers the presence of a chemical compound in leaves acts as a deterrent but for a few species it is an attractant. This explains why most species of plant-feeding insects are restricted to a relatively narrow range of species of plants. Co-evolution may be viewed in terms of adaptations by plants that inhibit damage from consumers and by consumers that enable exploitation of plants. There is, however, yet another possibility even more theoretical.

Caterpillars that feed on leaves and Homoptera that suck sugar and amino acids in solution from the phloem or xylem of leaves and stems produce copious quantities of waste material which of course falls to the ground beneath the plant. This material is decomposed and broken down into simpler compounds and it must be assumed that some of the nutrients that are released re-enter the plant in solution via the roots. Even in the most uniform tropical environment there are seasonal cycles of flowering and seed production, and it is possible that, on a seasonal basis, plants make use of their consumers as a means of concentrating nutrients in their tissues at certain times of the year. Seed production, for example, requires protein which the plant must manufacture, and to do this the plant needs nitrogen which may be in short supply. But nitrogen is present in the waste products of consumers which after decomposition can re-enter the plant. If, for instance, a tree produces leaves at a certain season and these are attacked by caterpillars, the waste products of the caterpillars may be available in greater concentration later in the season when seed is being formed and when nitrogen is most in demand. Thus it is possible that plants 'solicit' consumers at a certain season in order to maximize and concentrate seed production in another season. Nitrogen is only one possible requirement; there are many others, even including water. Thus in West Africa a large bug, *Ptyelus grossus*, extracts water (and presumably amino acids) from the xylem of certain species of trees. This it does in the dry season and trees heavily infected with bugs produce large quantities of water which falls to the ground; water is thus made available 'out of season'. Roots near the surface could then absorb nutrients from the soil in the dry season, and it may be, in this instance, that the bugs contribute to maximizing seed production later in the year, or even in the following year. It is possible that trees affected by the bug compete successfully with unaffected trees for soil nutrients.

It is easy to form the impression that consumers are invariably damaging to plants and the suggestion that they may have the opposite effect may be viewed with suspicion. The difficulty, I suggest, is that many ecologists are preoccupied with the effect of consumers on productivity: there is abundant evidence that consumers restrict productivity, but from an evolutionary point of view it is not so much productivity but seed formation that is important. No doubt the two are often correlated but if to form seeds there are special short-term nutrient requirements, one way of satisfying these is the evolution of a system that concentrates nutrients on a seasonal basis. It is possible to imagine, therefore, that trees and plants control the abundance of their consumers at a (normally) low level but at the same time allow a certain amount of consumption on a seasonal basis that facilitates rapid recycling of essential nutrients.

I mentioned earlier that these nutrients originate from the waste products of the consumers. There is one further possibility: leaves damaged by insect larvae often fall to the ground in advance of undamaged leaves, and hence are decomposed and their nutrients released earlier than would otherwise occur. The leaves of the umbrella tree, *Musanga cecropioides*, (Fig. 4.3) are eaten by the gregarious larvae of the butterfly, *Acraea pentapolis*. Chewed and damaged leaves drop from the tree out of sequence and earlier than undamaged leaves. They are therefore decomposed earlier and 'out of season'. Does this result in a more rapid recycling of nutrients? Fruit production in *Musanga cecropioides* is seasonal and it is possible that partial defoliation initiated by butterfly larvae results in a temporary seasonal increase in soil nutrients which could be used for fruit formation later in the year, or indeed in the following year.

In tropical forests it is unusual for animal populations to increase suddenly in size and become

Fig. 4.3 An umbrella tree, *Musanga cecropioides*, partly defoliated by the larvae of the butterfly, *Acraea pentapolis*. (Photographed in Sierra Leone by T. L. Green.)

'out of balance.' In savanna, too, there are rather few species that periodically increase in numbers, although there are well-known exceptions like locusts. In general the size of each population remains more or less constant year after year; there are no significant outbreaks and there is a distinct impression of long-term stability. Natural ecosystems at high latitudes are far more subject to large oscillations in population size, as are ecosystems that have been extensively modified by man. The problem of pests arises when natural vegetation is cleared and the land converted into a monoculture; under these conditions there is a substantial reduction in animal species diversity and an increase in abundance of those species that are able to exploit the changed circumstances.

I have thus far characterized ecosystems in terms of energy flow and nutrient cycling; I have also drawn attention to the apparent under-exploitation of terrestrial producers by primary consumers and developed a theory as to how this has been evolved. Ecosystems can also be characterized in terms of the diversity of species. All tropical areas, but especially forests, contain a much greater diversity of plants and animals than comparable temperate areas. It is tempting to suggest that the high diversity of species present in tropical ecosystems results in an increase in interactions between individuals to such an extent that it is impossible for the population of any one species to increase suddenly. In other words, high levels of predation and parasitism and intense competition for resources tend to maintain populations at stable equilibrium. This leads to the generalization that there is a correlation between species diversity and population stability, the most stable populations occurring in the presence of the greatest diversity. Evidence in support of this suggestion is largely circumstantial and comes mainly from what happens to ecosystems when they are disrupted by human activities. If a patch of tropical forest is cut down and the land planted with a single species of crop there is inevitably an outbreak of pests – nearly always insects. The pests may move in from elsewhere, or, in some instances, may be species that already occur in the area but which until the creation of the monoculture had remained relatively rare. Indeed the introduction of European-style monocultures into Africa has not always been successful, and it is interesting to note that people growing crops for their own use tend to plant a diversity of species within a small area. Perhaps repeated experience has taught them that outbreaks of pests are far less likely when plants are grown as mixed stands than when they are grown as monocultures.

All the individual trees of a species tend to produce seed at the same time. Considerable quantities of seed are eaten by animals, often before it falls to the ground and also when it is on the ground. Animals frequently eat (or digest) only the outer covering of the seed. They therefore play a part in seed dispersal. The number of seeds that eventually germinate beneath a tree is small compared to the number produced by the tree. In West African forests the animals mainly responsible for eating seeds include species that are extensively hunted for food. One effect of this is that with the decrease in numbers of the seed-eating animals there is nowadays a high rate of germination beneath the parent tree. Few seedlings attain maturity but, more importantly, there is probably much less seed dispersal than there would be if the seed-eating animals were still common. Thus hunting pressures by man can be expected to change to some extent the distribution and diversity of forest trees, and these in turn must affect the abundance of animals that feed on the leaves of seedling trees.

There are few entirely natural ecosystems remaining in tropical Africa. Even high on tropical mountains signs of air-borne pollution can be detected, and in most areas the impact of man is immediately apparent. Many of the forests are now reduced to tiny patches and the savanna is almost everywhere heavily grazed by domestic animals. The conversion of the African forest and savanna into cultivated and grazing land has produced a new type of ecosystem, perhaps equally complex, and certainly as little understood as the ecosystems that have so recently been destroyed. This new ecosystem is characterized by a mixture of forest and savanna plants and animals, and by cultivated plants and weeds and the animals that have been able to exploit them.

Lastly, consider the different consequences of cutting down a tropical forest and a northern coniferous forest. Neglect for a moment the difference in species diversity and its possible effects on population stability in the two ecosystems and concentrate instead on what happens to the nutrients. Both ecosystems contain about the same amount of organic carbon. In the coniferous forest about half is in the soil while in the tropical forest about three-quarters is in the wood. The difference arises because in the temperate region decomposition is slow while in the tropics it is extremely rapid. Most of the organic carbon (and indeed other nutrient material) in a tropical forest is locked up in living vegetation. The implication of this difference is far-reaching and important. If a tropical forest is cut down and the timber removed most of the nutrients are lost, but if a coniferous forest is cut down a substantial proportion of the nutrients remain in the soil. Thus in temperate areas soil left after the removal of trees is quite fertile, but in a tropical area the soil remaining is almost devoid of nutrients. This explains why cultivation does not always succeed in apparently rich tropical areas. Perhaps the lesson here is that the more complex ecosystems, although stable, are the most delicately balanced and that there are greater repercussions when they are altered or destroyed.

Chapter 5

The seasons and other periodic events

In the temperate regions most biological events are seasonal. In Britain, for instance, birds breed chiefly in April, May, and June; each species of butterfly flies at only a certain time of year; many invertebrates and some vertebrates hibernate; and many insects and birds migrate, arriving from the south in the spring and departing for the south in the autumn. These events are regular and predictable: if an insect flies as an adult in a particular month in one year it will normally do so every year.

Such regular seasonal events are obviously adapted to the occurrence of a clearly defined winter and summer and a somewhat less clearly defined spring and autumn. The main environmental factors that determine seasonal biological events in temperate regions are incremental changes in daylength and changes in air temperature. Changes in daylength are absolutely regular and predictable and it is therefore not surprising that a great many animals, particularly birds and insects, respond readily to such changes. Temperature fluctuations, although regular, are not absolutely predictable; everyone living in the temperate regions is familiar with the occasional mild or very cold winter.

A visitor from the temperate regions is perhaps at first surprised to find that in the tropics, even near the Equator where seasonal changes in daylength and temperature are small, most animals have a seasonal rhythm in many of their activities. In the lowland tropics the seasons are less obvious to the casual observer, for a number of different reasons. First, especially in forested regions, the habitat looks uniformly green all the year round. Secondly, whereas in temperate regions a whole group of animals (such as birds) may be more or less synchronized in such activities as breeding and migrating; in the tropics they are not. In Europe most birds breed in the spring, but in a tropical country such as Uganda birds may be found breeding in all months of the year, although for a particular species breeding may be restricted to a relatively short period. Thirdly, for reasons that follow, many seasonal events that occur once a year in the temperate regions may occur twice a year in the tropics.

In the tropical regions of the world, including tropical Africa, there are more or less predictable seasonal changes in the amount of rainfall. In many, but not all, tropical areas there are two relatively wet and two relatively dry seasons of the year. The word 'relatively' is important because in some regions, especially in the equatorial forests, rain may fall all the year round and it is only by careful measurement over several years that a seasonal pattern emerges. In the absence of marked changes in daylength and temperature it should come as no surprise that in tropical Africa most seasonal events are associated directly or indirectly with seasonal fluctuations in the rains. In the temperate regions rainfall may on occasion modify the effects of changing daylength and temperature, but in the tropics the situation is reversed: temperature and daylength may on occasion modify the effects of rain.

Proximate and ultimate factors

In discussing seasonal and other rhythmic events, it is convenient to distinguish two kinds of factors that are important in maintaining the rhythm. These may be referred to as proximate and ultimate factors. Proximate factors are environmental events that serve as a trigger to an animal's physiology; they convey to the animal information about the time of year, time of day, likelihood of food shortage, and so on. The many ways in which proximate factors operate present problems for the environmental physiologist rather than for the ecologist. Proximate factors stimulate a rhythm but are not themselves the ecological reason for the rhythm; their effect is to synchronize events for a given population. Ultimate factors, on the other hand, are events that determine why an animal breeds, migrates, moults, aestivates, and so on, at a particular time. Presenting problems for the ecologist and evolutionist rather than for the physiologist, ultimate factors are periodic environmental events that, for a particular population, make resources more available at certain times than at others. Here is a classification of some proximate and ultimate factors that are of importance to animals in tropical Africa:

Proximate factors (or how the animal knows when to respond): changing length of day; fluctuation in air temperature; seasonal rainfall; variations in relative humidity; diurnal fluctuations in amount of illumination; the phases of the moon.

Ultimate factors (or why the animal responds at a particular time): availability of food; availability of breeding site; absence or comparative rarity of predators and parasites; synchronization of appearance of breeding adults; seasonal availability of water and humid conditions; seasonal fires and seasonal drought.

A proximate factor can under certain conditions also act as an ultimate factor. In frogs, for example, seasonal rainfall can stimulate breeding proximately and can at the same time be the ultimate reason for breeding at a particular time. Very often, though, proximate and ultimate factors controlling the occurrence of a biological rhythm are different. Thus in Africa seasonal rainfall often acts as a proximate stimulus for birds to breed but the ultimate factor is increased availability of food.

Exogenous and endogenous rhythms

Most biological rhythms in animals appear to be controlled by environmental events acting in the first place through the nervous and hormonal systems. Environmentally determined rhythms are usually known as exogenous. But there are also rhythmic events that cannot be easily correlated with environmental stimuli, and it is postulated in such cases that there is a rhythm within the animal itself; this is called an endogenous rhythm. It is difficult to say how widespread endogenous rhythms are, because although it is possible to show that exogenous rhythms occur it is impossible to show that they do not occur. Hence, to postulate that there is an endogenous rhythm, it is first of all necessary to eliminate the possibility of occurrence of rhythmic external events, an extremely difficult task. In some well-investigated examples it appears that exogenous and endogenous rhythms can reinforce one another. Near the Equator, where changes in daylength and temperature are small, it is tempting to suppose that endogenous rhythms are relatively more important than in temperate regions. Whether this is so remains to be investigated. My impression is that the flowering pattern of many tropical plants is such that endogenous rhythms may be involved. The flowering pattern of a single iris plant, *Marica caerulea*, in my garden at Kampala is perhaps suggestive of an endogenous rhythm. The flowers of this plant last only one day, and flowering is cyclic. Several flowers (once there were 52) appear on one day and then there are no flowers for an interval of a few days. During 105 consecutive days in 1964, a mean of 3·5 flowers occurred once every 3·1 days. Hence the flowering of this plant is neither random nor seasonal, nor does it appear to be associated with any climatic or other external factor. It is possible, then, that in this plant there is an endogenous rhythm the overall effect of which is to synchronize periodic flowering.

The development of a bright yellow breeding plumage in a small passerine bird, the yellow wagtail, *Motacilla flava*, just before departure from its equatorial 'wintering' grounds for the breeding grounds in northern Europe and Asia has been cited as an example of an endogenous rhythm (Marshall and Williams, 1959). These birds, along with many other migratory species, leave their breeding grounds in Europe and Asia in late summer and autumn under the proximate stimulus of decreasing daylength. They migrate to equatorial Africa, arriving in late September, and depart again for the north in March and April. In southern Uganda in January many of the birds begin to develop the bright yellow plumage associated with the breeding season. This area is 0°04′N. and changes in daylength are very small; temperature fluctuations are irregular and slight, and in every month there is usually more than 6 cm of rain. Marshall and Williams (1959) claim that there is no seasonal change in the supply of insect food, and postulate because of this that an endogenous rhythm is at least partly responsible for the development of bright plumage in January before the commencement of the return flight in April. In fact there is ample evidence of considerable fluctuations in the numbers of insects near the Equator; indeed in southern Uganda the seasonal peak of insect abundance coincides with the departure of the yellow wagtail (Owen, 1969) and it is likely that this is the main proximate stimulus involved. Long-distance migrants like the yellow wagtail lay down large amounts of fat before they migrate. The fat provides the fuel for the long northward journey, and in order to fatten the birds undoubtedly require a good supply of food.

Much more convincing evidence of endogenous rhythms comes from studies of the egg-laying cycles in the mosquito, *Aedes aegypti* (Gillett, 1965). In this species individuals exposed to varying lengths of artificial light and dark lay eggs at a specific time in a 24-hour period, and the cyclic activity occurs even when they are exposed to only 5 minutes of light a day.

The seasons at Kampala

Kampala, at 0°20′N., is almost on the Equator. There are about 12 hours of daylight and 12 hours of darkness all the year round. The mean maximum temperature is around 80°F. (about 27°C.) and the mean minimum is 60°F. (about 16°C.). Fluctuations in temperature in each 24-hour period are greater than seasonal changes. There is virtually no seasonal change in temperature: January and February are on average a little warmer than June and July, but the difference is not obvious. Minimum temperatures normally occur at night and maximum in the afternoon. Once or twice a month, after heavy rain, the temperature may fall as low as 55°F. (about 13°C.). The Kampala sky is rarely completely clear and it is often very cloudy, especially in the afternoon. Thunderstorms are frequent in the early afternoon or during the early hours of the morning, usually coming from Lake Victoria to the south. Some storms may be accompanied by hail and high winds which cause considerable damage to crops. Prolonged light rain may occur in the normally wetter months of April and November but most of the rain falls during thunderstorms. The vegetation is generally lush all the year round, and superficially there is an impression that most animals are performing most of their activities regardless of the time of year. Many animals, especially insects and birds, are responsive to rainfall and activities may start and stop with the onset of rain. The monthly rainfall at Kampala for the 10 years 1953–62 is shown in Table 5.1. During this period the total annual rainfall varied between 35 and 64 in (889 and 1625 mm) with a mean of 46 in (1168 mm). The mean monthly rainfall is always in excess of 2 in (51 mm) and never over 7 in (178 mm), July being on average the driest month and April the wettest. On average there are two relatively wetter seasons, March–April and October–November, and two relatively drier seasons, January–February and June–July. But the means are deceptive, as shown in Table 5.1. The wettest month of the year may come earlier or later than usual in a particular year and sometimes, as in 1954, a normally dry month (July) may be wet while a normally wet month (November) may be dry. Many animals living in the

Table 5.1 *Monthly rainfall at Kampala, Uganda, 1953–62 (in inches). Kampala is at 0°20′N. (Figures supplied by the East African Meteorological Department, Entebbe.) (1 inch = 25·4 mm)*

Year	Jan.	Feb.	Mar.	Apr.	May	June	July	Aug.	Sept.	Oct.	Nov.	Dec.	Total for year
1953	2·77	1·84	1·79	4·98	2·77	5·94	0·72	3·43	5·31	7·31	4·46	2·04	43·36
1954	0·32	2·67	2·10	2·53	5·75	2·30	4·95	3·37	3·74	2·69	0·89	3·73	35·04
1955	3·40	5·54	6·31	7·92	5·31	2·21	3·04	6·72	4·64	2·11	2·37	3·91	53·56
1956	4·03	0·62	2·51	9·46	2·29	0·91	0·52	3·74	1·87	2·14	5·41	3·71	37·21
1957	3·52	1·71	8·02	8·34	0·73	2·85	1·89	2·88	1·43	2·27	3·58	5·48	42·70
1958	1·70	2·60	6·18	6·65	3·71	2·74	2·94	3·44	4·60	2·91	3·58	5·01	46·06
1959	1·84	1·50	3·26	4·67	1·66	0·78	1·60	7·08	3·93	3·83	8·19	3·34	41·68
1960	5·11	2·03	6·91	9·30	6·90	1·86	1·93	2·37	4·48	4·01	2·06	1·24	48·20
1961	0·00	1·62	9·01	7·27	5·00	1·69	1·90	5·77	4·93	12·77	9·93	4·52	64·41
1962	2·41	0·92	8·58	7·30	3·31	2·78	0·65	4·84	3·47	7·61	4·60	4·74	51·21
Mean	2·51	2·10	5·47	6·84	3·74	2·41	2·01	4·36	3·84	4·77	4·51	3·77	46·34

Kampala area exhibit some fluctuations in their activities and in most cases rainfall appears to be the most important proximate factor. For instance, every time there is heavy rain or even the likelihood of heavy rain, a small bird, the white-browed robin-chat, *Cossypha heuglini*, bursts into melodious song. This bird rarely sings unless it rains. Mole crickets, *Gryllotalpa* (Fig. 5.1),

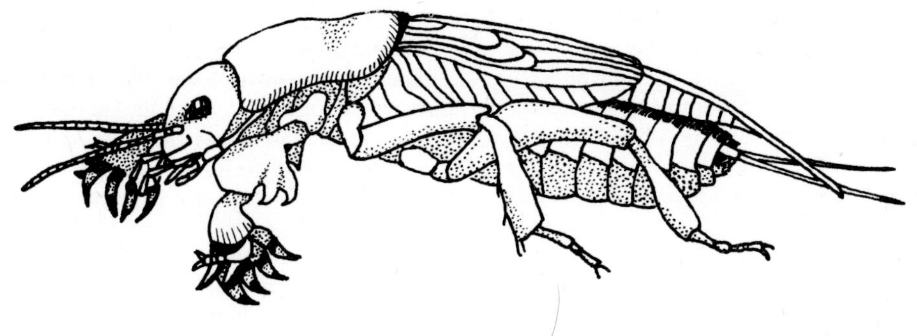

Fig. 5.1 A mole cricket, *Gryllotalpa* sp.

also strongly influenced by rainfall will sing after rain, whereas during dry spells they are hardly ever heard. Winged termites and ants take flight soon after rain. Indeed the biological activity of many animals around Kampala is strongly influenced by rain, far more so than any other single environmental event.

Much the same situation prevails elsewhere in Africa, the main variable being the amount, seasonality, and predictability of rainfall. In Sierra Leone, for example, there is three to five times the rainfall that occurs at Kampala. But the Sierra Leone rainfall is more seasonal and predictable, every year having a well-defined dry season with no rain and a very wet rainy season with up to 50 in (1270 mm) falling in the wettest month. And, as would be expected, animal activity is altogether more seasonal in Sierre Leone than at Kampala: there is, for example, an enormous increase in numbers of flying insects at the onset of the rainy season in April and May.

The breeding seasons of animals in tropical Africa

Birds are the only major group of animals in tropical Africa for which there is a reasonably complete picture of breeding seasons. For most species rainfall appears to be the most important proximate stimulus to breed. In East Africa, away from the influence of the large lakes, there are two relatively well-defined rainy seasons in the year and breeding tends to occur twice a year, either in the dry seasons or in the wet seasons, depending upon the species. The most important ultimate factor affecting the breeding seasons of birds appears to be seasonal availability of food for the young but few species have been studied in detail from this point of view. Many birds feed on insects and everywhere in Africa where measurements have been made there is at least one seasonal peak of insect abundance. Chapin (1932) has discussed the breeding seasons of birds in the lowland equatorial forest of Zaïre. Here rainfall is well distributed and temperature fluctuations are small. The breeding activity of passerine birds is less in the drier months of November–January but there are many species that are exceptions. Woodpeckers breed especially in November–January and doves and birds of prey tend to avoid the wetter months of February–May.

It is, however, possible that there are seasonal peaks in activity even in areas of equatorial Africa where birds of a particular species breed in every month of the year. For instance a colony of black-headed herons, *Ardea melanocephala*, at Nairobi, Kenya, was occupied by breeding herons in every month for the 4 years 1958–62. Indeed there is evidence that a colony of this species has been used continuously for 20 years at Kampala (North, 1963). But although the black-headed herons at Nairobi breed continuously, their activity varies; in particular the number of birds breeding declines during dry weather and increases with the rains.

The breeding seasons of mammals are not as well documented as those of birds, although there is at present considerable interest in the subject. In the savanna an important ultimate factor for ungulates may be the appearance of new vegetation (especially grass) after the rains, and climatic variation may be, as in birds, the most important proximate factor. The availability, during and after the rains, of long grass in which new-born young can hide may also be significant. Some carnivores breed when their chief prey species are themselves giving birth to young. Many mammals in tropical Africa appear to breed all the year round, but in the rather few instances where large samples have been examined, distinct peaks in breeding activity have been found.

Kellas (1955) examined a sample of 412 specimens of a small antelope, *Rhynchotragus kirkii*, collected at Shinyanga, Tanzania (3°33′S.). Females were found to be polyoestrus throughout the year, with two peaks of intensive breeding, one at the beginning of the rains in November and one at the end in April. The gestation period in this species lasts about 6 months and most females become pregnant twice a year. Spermatozoa occur in adult males at all times of the year but the testes show a seasonal fluctuation in weight that corresponds with the breeding cycle of the females. Young born at the beginning of the rains are able to enjoy the cool long-grass season, while those born at the end of the rains spend their first few months of life in hot, dry conditions during which the vegetation becomes more and more sparse. The two seasons into which the young are born are hence very different, but both lots of young are dropped when the grass is long; this, then, may be the most important ultimate factor.

What little information there is for the primates suggests that many species breed all the year round. In Uganda the red-tailed monkey, *Cercopithecus ascanius* (*nictitans*), appears to breed seasonally as shown by the following records (Table 5.2) collected by Haddow (1952). The

Table 5.2 *Seasonal breeding of Cercopithecus ascanius (nictitans.)*

Period	*Per cent in early pregnancy*	*Per cent in late pregnancy*	*Per cent with infants*
January–April	19	—	31
May–August	4	36	—
September–December	—	9	18

sample of 52 upon which the table is based is admittedly a small one, but there appears to be a peak of breeding activity in December–April. Other species of *Cercopithecus* (common monkeys throughout Africa) may be seasonal in some areas and non-seasonal in others.

Mutere (1965, 1967) has found that in the fruit bat, *Eidolon helvum*, living near the Equator at Kampala, not only is breeding strongly seasonal but there is also delayed implantation. The young are born in February and March. The weight of the testes varies seasonally, reaching a peak in April–June, when also spermatozoa are most abundant. Spermatozoa are found in the genital tracts of the females only during April–June but implanted embryos do not occur until October. During July–September unimplanted blastocysts are present in the uterus and there is hence a delay of about 3 months between fertilization and implantation. This appears to be the first record of delayed implantation in a tropical mammal and in a bat; delayed implantation has until now been known only from mammals in temperate regions. In *Eidolon helvum* the ecological significance of delayed implantation, and indeed of seasonal breeding, is obscure; the bats feed largely on cultivated fruits which do not appear to vary in abundance seasonally.

In The Gambia the lung-fish, *Protopterus annectens*, aestivates when seasonal drying up of the swamps commences in November and December. Lung-fish remain in aestivation for 7 or 8 months until the rains and then, once the swamps are flooded, they begin to breed. Breeding is thus strongly seasonal and may be earlier or later in any particular year, depending on the condition of the swamps (Johnels and Svensson, 1954). Less is known about the breeding of the East African lung-fish, *Protopterus aethiopicus*: in Uganda breeding occurs chiefly in November–April and, as in *Protopterus annectens*, is associated with higher rainfall and local flooding; unseasonal breeding may occur if there are unseasonal rains (Greenwood, 1958).

In Lake Malawi the rock-frequenting cichlid fish seem to breed all the year round (Fryer, 1959). In northern Lake Victoria the cichlid, *Tilapia variabilis*, also breeds all the year round, but with peaks at somewhat irregular intervals especially in November and December and in July (Fryer, 1961). The seasonal breeding patterns of lake and river fish in tropical Africa pose a number of interesting questions: a great many deep-water species move into shallower water for spawning but the proximate factors stimulating this movement are not known. Anadromesis, or up-river migration, occurs in a number of species. At certain times of the year (especially during the rains) large numbers of several species of Cyprinidae and Clariidae migrate up-river from Lake Victoria. Breeding then takes place in the swampy upper reaches of the rivers. Little is known of either the proximate or the ultimate factors involved in anadromesis; rising water levels seem important and it is thought that the migration ensures better feeding grounds for the young fish and reduces the chance of predation from larger lake fish.

The entire activity, including breeding, of tropical African frogs and toads is much influenced by rain. Like birds, many frogs begin to sing and call soon after rain has fallen and often frog spawn may be found in temporary puddles formed after a downpour. Many of the smaller forest tree-frogs breed in seasonal water-holes in trees. One of the most widespread African amphibians, the toad, *Bufo regularis*, has almost everywhere a peak of breeding in the wet season, but in Sierra Leone it breeds in the dry season (Menzies, 1963).

In most areas where it still occurs the crocodile, *Crocodilus niloticus*, lays its eggs during the dry season and hatching occurs with the onset of the rains. In Uganda the crocodile occurs in the Nile and from Lake Kyoga northwards breeding takes place once a year in December and January. But to the south, in northern Lake Victoria, breeding occurs twice a year: in August and September and in December and January (Cott, 1961). In other parts of Africa there is evidence that crocodiles lay when the weather is dry and the young appear with the rains. In Ghana the largely insectivorous rainbow lizard, *Agama agama*, breeds when the weather is wet, possibly because this is when food is most abundant (Chapman and Chapman, 1964).

Figure 5.2 shows the seasonal pattern of breeding of the land snail, *Limicolaria martensiana*, at

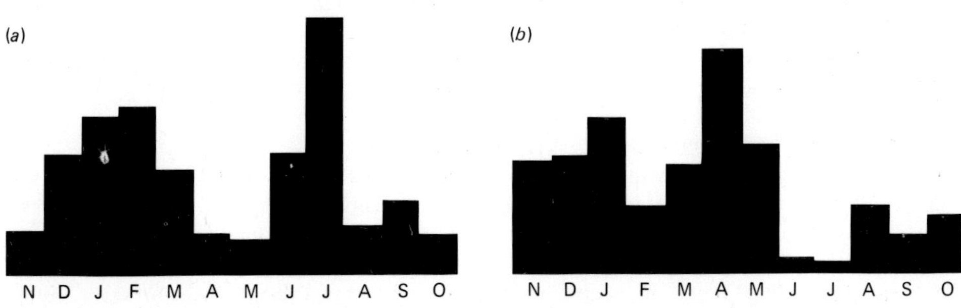

Fig. 5.2 The breeding season of the land snail, *Limicolaria martensiana*: *(a)* Percentage of snails with eggs based on a sample of 5318 snails, about 450 being examined in each month; *(b)* Rainfall (inches). (From Owen, 1964.)

Kampala. Weekly samples of snails were collected from a single population from November 1962 until October 1963. After collection the snails were removed from their shells and the number with eggs recorded. The upper histogram in Fig. 5.2 shows the percentage of adult snails with eggs collected in each month from November until October. The snails produce eggs all the year round but there are two peaks, one in February and the other in July (when 33·1 per cent of the snails contained eggs). These peaks correspond with the normally drier months of the year. In May only 4·8 per cent of the snails contained eggs. The lower histogram in the Fig. shows the actual monthly rainfall during the period of collection – April being the wettest month with 9·4 in and July the driest with 0·4 in. Note that during this period the dry season was drier than usual and the wet season somewhat wetter – see Table 5.1. The seasonal pattern of breeding found in *L. martensiana* would result in many newly-hatched snails feeding during the wettest months of the year when suitable vegetation for food would be more available and where there would be less risk of desiccation.

This and other populations of *Limicolaria martensiana* have now been sampled for several consecutive years and the seasonal pattern of breeding established for one year has been confirmed but with variations that can be attributed to variations in rainfall (Owiny, 1974). In *L. martensiana* the alternation of relatively wet and relatively dry seasons could provide the proximate stimulus for fluctuations in breeding activity, and the chief ultimate factor is the increased general suitability of the habitat when it is wet. In the Western Rift of Uganda the same species of land snail is more seasonal in its breeding; this is presumably because the wet and dry seasons are much more sharply defined and regular. *Limicolaria martensiana*, then, resembles the crocodile in its breeding pattern: both species lay eggs mainly when it is dry and the young appear when it is wet.

Rainfall has a marked effect on the swarming pattern and seasonal colony-formation of termites. New colonies of termites are founded by the simultaneous emission of winged reproductive males and females from established colonies. In many species the proximate stimulus for synchronized emission is rainfall after a dry spell. Often winged reproductives accumulate for a time before setting out on their colonizing flight. During this period the gonads are not fully mature and there is no sexual behaviour. Just before the flight from the colony the worker termites make exit holes in the walls of the termite mound or burrow. Numerous workers and soldiers congregate around the exit holes when the reproductives fly off. Males and females leave the colonies in about equal numbers. They fly relatively short distances, mate and found new colonies. Many species of African termites are seasonal in their colony-founding flights and most species are stimulated to fly by rainfall. Presumably rain is also the ultimate factor in termite flights: the ground would be softer and more suitable for starting a new mound or burrow after rain. The sudden appearance of vast swarms of reproductives attracts numerous predators, including people. In the Ivory Coast more than 150 species of birds have been recorded as exploiting termite swarms, and interestingly enough each species of bird hunts termites in a characteristic way; it is as if the ecological segregation between species persists even in the presence of sudden super-abundance of food (Thiollay, 1970).

Migration and seasonal movements

Migration is a regular movement to and from a breeding area. In birds and other vertebrates the same individual may take part in both movements, in insects different individuals usually take part in the two movements. Migration, then, is a regular shift of a population from one place to another for the purpose of breeding.

Many birds that breed in temperate Europe and Asia spend the northern winter (September–March) in tropical Africa. Some Arctic species – including tiny warblers – migrate south, crossing Europe, the Mediterranean, and the Sahara to spend the northern winter in Africa. Many of the migrants are insectivorous species such as warblers, flycatchers, swallows, and swifts; others are birds that breed on tundra and moorland and migrate south to spend the winter around the shores of tropical lakes and ponds. The proximate stimulus for the departure of migrant birds for Africa is in many species decreasing daylength in the autumn, modified by decreasing air temperature. The chief ultimate factor is decreasing availability of (especially insect) food, and perhaps also cold weather. It is not completely understood how the migrants know when to leave tropical Africa and return to the north: some favour the existence of an endogenous rhythm, others feel that changes in the food supply act as the trigger for departure. But why do so many birds undertake such extensive migrations? The answer seems to be associated with variations in food supply. At high latitudes insects in particular become extremely abundant for a short period of the year, far more abundant than in most tropical areas, and it must be assumed that migration is an adaptation to this regular seasonal increase in the food supply.

It was at one time believed that northern birds reached tropical Africa by specific routes, the West coast and the Nile often being cited in this connection. But it is now known that many species can make a direct crossing of the Sahara without refuelling and that migration is often on a broad front and not along narrow routes.

Tropical Africa also receives migrants from South Africa and from Madagascar, and there are many species that migrate within the tropics. In general, the birds that breed in the equatorial forest are less migratory than those that breed in the savanna, presumably because seasonal climatic changes are less marked in the forest than in the savanna. The weaver birds are an abundant family in Africa, occurring both in the forest and in the savanna. Most forest species are non-migratory but some savanna species undertake short-distance migrations. The pennant-winged nightjar, *Semeiophorus vexillarius*, is migratory within tropical Africa. It has a wide breeding area south of the Equator and migrates north during the southern 'winter', though still remains in the tropics, as shown in Fig. 5.3.

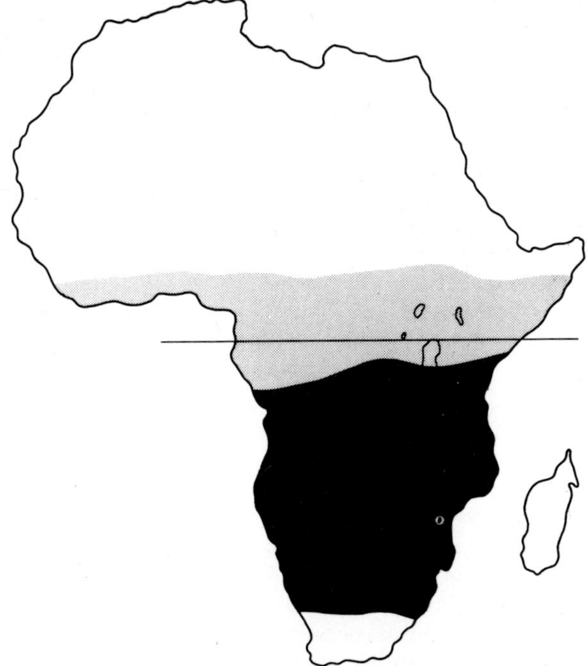

Fig. 5.3 The breeding area (solid black) and migration area (shaded) of the pennant-winged nightjar. *Semeiophorus vexillarius*. (Redrawn in part from Mackworth-Praed and Grant, 1952.)

In savanna areas where there is seasonal drought there may be considerable migrations of large mammals. The elephant and the wildebeest are mammals that in parts of their range in East Africa undergo regular migrations, though less so than in former times because of pressures exerted by expanding human populations. Seasonal shortage of vegetation for food and of water are probably the ultimate factors involved in these movements.

Within tropical Africa many insects are migratory – nearly all of them savanna species – but little is known of the periodicity of the movements or of the ecological factors causing the migrations. Butterflies of the family Pieridae have often been encountered moving in fixed directions, sometimes in immense numbers. In West Africa the butterfly, *Libythea labdaca*, migrates in a general southerly direction in February–May and in a general northerly direction in October–December (Williams, 1951). Observers throughout Africa have repeatedly reported the sudden occurrence of immense swarms of this species but where they have come from and where they go remains a mystery. In the more humid parts of East Africa the long-horn grasshopper, *Homorocoryphus nitidulus*, migrates in swarms at particular times of the year. This species is closely associated with the rains of November and April, and in southern Uganda swarms normally occur in November and December and in April and May. *Homorocoryphus nitidulus* is nocturnal and is attracted to street lamps at night. It is extremely fat, a feature of many insects and birds that migrate long distances. In West Africa it is apparently non-migratory.

Irruptions

Irruptions are irregular non-seasonal movements, usually caused by a failure of local resources – especially food – or by high population, or by both. Irruptions are a feature of the climatically less stable Arctic and sub-Arctic areas of the world; the periodic mass movements of lemmings and certain species of seed-eating birds are examples of irruptions.

In Africa the desert locust, *Schistocerca gregaria*, is one of the most conspicuous animals that undergoes irregular mass movements. This species is a potential pest over a vast area of Africa and tropical Asia, affecting about 60 countries. Within this area more than 300 million people, one-eighth of the world's population, are liable to suffer devastation of their crops. Numbers of locusts may be so low for a number of years that it may even be difficult to find specimens for experimental work. Then there are sudden and unpredictable increases in small local populations, followed by mass irruption. It is then that damage to crops occurs. The desert locust feeds on a wide variety of crops and other plants and can cause almost complete defoliation during an irruption. Swarms (Fig. 5.4) comprise up to 1 000 million individuals and can consume 3 000 tons of food a day. Figure 5.5 is a map showing the area of Africa liable to invasion by the desert locust. At present, possibly because of control measures, it is exceptional for swarms to reach the equatorial parts of Africa, but areas of northern Kenya and the southern Sudan are likely to suffer periodically. The nymphs of the desert locust are green when not at high density but as population density rises they become darker and take on a distinct yellow and black coloration. This transition, which is accompanied by many behavioural, physiological, and morphological changes, is referred to as the transition of phase *solitaria* to phase *gregaria*. There are many theories regarding the significance of phase transformation in this and other locusts but despite an enormous amount of experimental and field study the function of the phases is still not fully understood. On one point there is no doubt: rising population density is the most important proximate stimulus for transition from *solitaria* to *gregaria*.

I have discussed irruptions in this chapter because in many ways they can be considered as a special kind of seasonal movement. They seem to occur in areas where normal seasonal stimuli break down or where populations become exceptionally large.

Fig. 5.4 An irruption of the desert locust, *Schistocerca gregaria*. The picture shows part of a 130 km^2 swarm over the airport at Hargeisa, Somalia, August 1960. The aircraft is one used by the Desert Locust Survey for spraying and destroying swarms of locusts. (Photograph by A. J. Wood, Desert Locust Survey.)

Aestivation

One way to avoid a seasonally unfavourable tropical environment is to move away from it; another is to aestivate. Aestivation occurs in many invertebrates and in a few vertebrates that live in areas that become seasonally or periodically dry. Species that live in shallow lakes or in swamps that dry up seasonally often survive the dry spell by aestivation in the mud, soil, or among vegetation. Three examples of aestivation from different groups of animals are discussed below.

During the dry season the River Gambia is restricted to its bed and the surrounding low-lying land is dry. During the rains this land is flooded and becomes a vast swamp, some of which is cultivated for rice. The onset of the rains, as well as the amount of rain, varies from one year to another, and this has an effect upon the aestivation cycles of a variety of species of animals, including the lung-fish, *Protopterus annectens*. Aestivation in this species has been studied by Johnels and Svensson (1954) from whom the account that follows has been taken. As the swamps dry up the lung-fish prepare themselves burrows in the mud. Unlike most fish that are found in the swamps during the rains, they do not move into the river when the dry weather sets in; in fact lung-fish are hardly ever seen in the river itself. The local name of the lung-fish is 'cambona', and when fishermen are asked whether they catch cambona in the river, they reply, 'Cambona is not an ordinary fish. It does not follow the water but the water comes to cambona.'

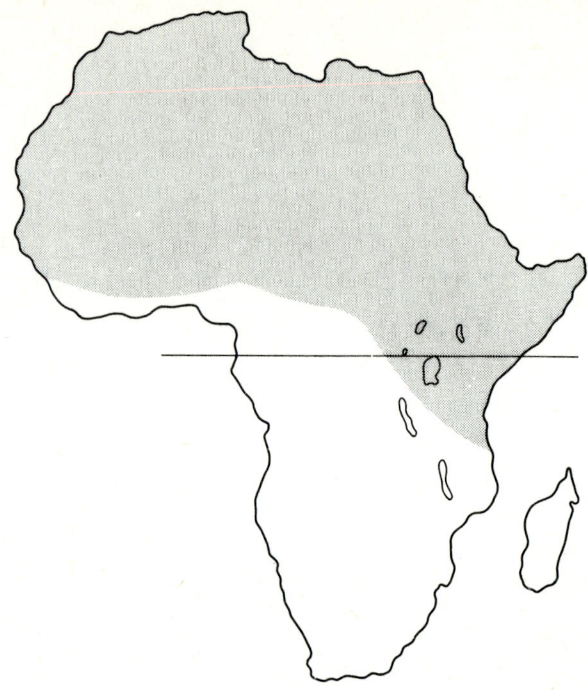

Fig. 5.5 The area of Africa that is subject to mass irruptions of the desert locust, *Schistocerca gregaria*. (From *The Hungry Thief*, published by Shell Limited, 1956.)

Figure 5.6 shows the structure of the completed aestivation burrow of a lung-fish. At the surface of the swamp above the burrow there is an earth lid (dotted in the Fig.); below the lid is a tubular channel, and finally there is the aestivating chamber of the lung-fish. The lung-fish fits tightly into this chamber and its normal position is shown in the Fig. Here it remains until the next rains. In some years when the dry season is wetter than usual many lung-fish do not complete the process of forming a proper aestivating chamber, but simply burrow into wet mud. The extent and length of aestivation appears to depend entirely on the dryness of the season. The aestivating burrow of the East African lung-fish, *Protopterus aethiopicus*, is quite different from that of *P. annectens*. It is a long chamber opening horizontally from a bank, and its base is filled with water (Wasawo, 1959). But rather few aestivating chambers of *P. aethiopicus* have been found and there may well be considerable variation in the mode of construction, depending upon the habitat and the extent of the drought.

During dry weather most species of land snails of the family Achatinidae retire into their shells and secrete over the aperture an epiphragm which effectively seals the snail into its own shell. The snail does not then move or feed and during dry weather is prevented from becoming desiccated. As soon as rain falls the epiphragm is shed and the snail becomes active. In forested areas where rainfall is frequent and there is no real seasonal drought, aestivation may last only a few days or at the most a few weeks, but in the seasonally drier savanna aestivation may last several weeks or even months. In the laboratory achatinid snails may be kept in aestivation for 3 or 4 months (some are reported to have been kept for 2 or 3 years); they become active immediately they are put into water. Aestivation is normally confined to dry periods but curiously enough there are nearly always a few snails in most populations in 'aestivation' even during the wet season. The

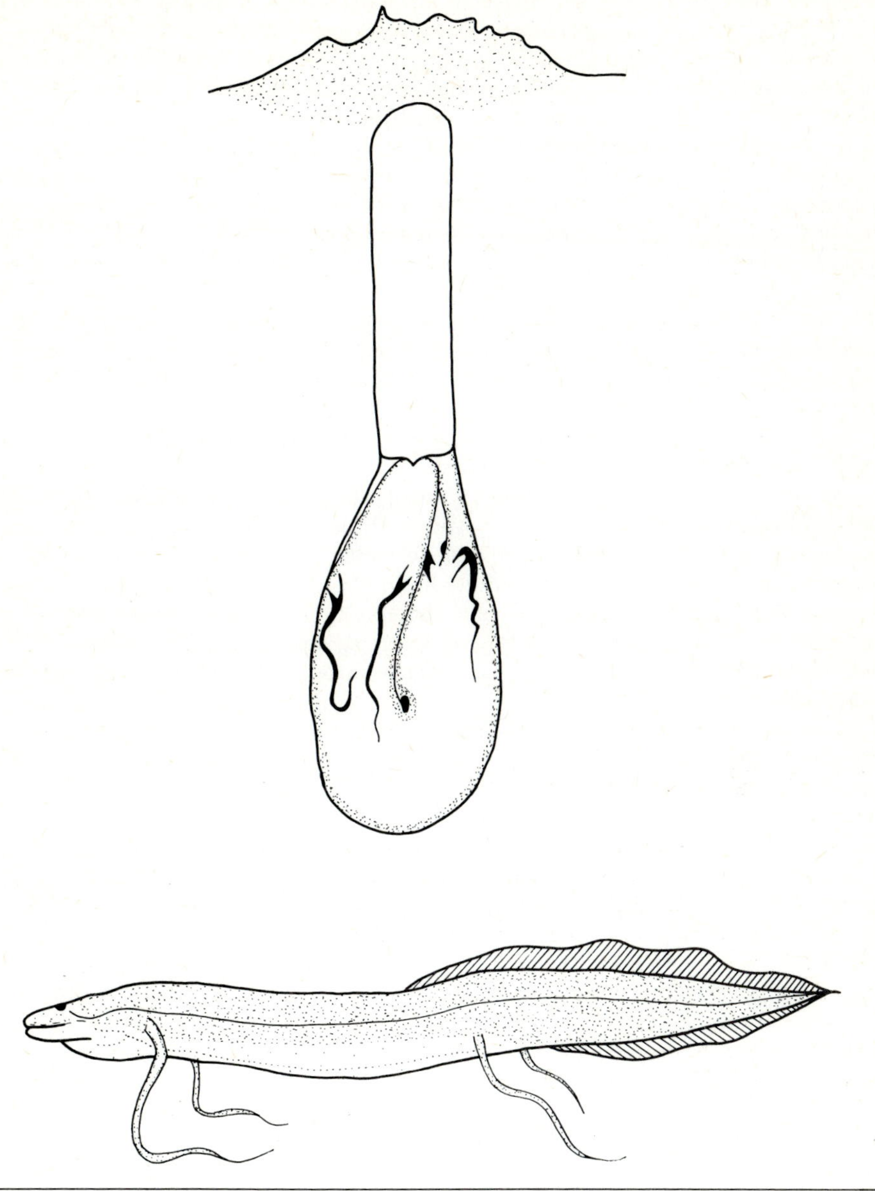

Fig. 5.6 An aestivation burrow of the lung-fish, *Protopterus annectens*. The lower figure is a lung-fish in the rainy season after aestivation. (Partly from Johnels and Svensson, 1954.)

significance of this is not known. Howes and Wells (1934) discussing water relations of snails and slugs in temperate regions, suggest that there is a tendency on the part of individuals for phases of aestivation to alternate with phases of activity even under approximately constant environmental conditions. They suggest a natural hydration cycle in the animals themselves. If this is so in the African Achatinidae, some of which burrow into the ground before secreting an epiphragm, catastrophic events such as grass fires, which often destroy great numbers of snails, would spare aestivating individuals. In species that feed on crops such as the giant snail, *Achatina fulica*, most attempts at mass destruction would still leave aestivating individuals alive and the population could then regenerate. The giant snail can in fact aestivate for 10 or 12 months, possibly longer (Mead, 1961). During this time the snails lose about 60 per cent of their body weight. Several organ systems are drawn upon for sustenance, including the reproductive system, which becomes atrophied so as to resemble superficially the juvenile condition. Aestivation in the giant snail ceases with the onset of rain. In *Limicolaria martensiana* fully formed eggs may remain viable for a month or more inside an aestivating snail. It is possible for an aestivating snail to shed its epiphragm without the stimulus of rain; direct sunlight will induce *L. martensiana* to shed its epiphragm and seek shelter, which, once found, enables the snail to enter aestivation again.

Snails of the family Achatinidae are, then, extremely sensitive to weather, and at any time after hatching they can enter aestivation as soon as the weather becomes dry. Since snails cannot move long distances away from poor conditions, aestivation becomes of vital importance to their survival.

There are many examples of aestivation in insects, especially in species living in seasonally dry areas. Insects may aestivate in any one of the stages in their life cycle from eggs to adults, but the stage at which aestivation normally occurs is usually characteristic of the species. In the Rukwa Valley, Tanzania, adults of the red locust, *Nomadacris septemfasciata*, emerge at the end of the rainy season in March and April. During the following 7 or 8 months until the next rains in October and November they do not breed and are, in effect, in a form of reproductive aestivation. Norris (1962) has shown that aestivation in this species can be induced in the laboratory by artificially changing the daylength; it therefore seems possible that changing daylength may be a proximate stimlus in natural situations. The ultimate factor involved in the aestivation of the red locust is, however, seasonal drought.

It is tempting to think of the dry season as the worst time of the year for animals and this is probably true in most instances. There are, however, species that encounter difficulties in the wet season, particularly in areas where rainfall is heavy. In Sierra Leone the pupae of swallowtail butterflies slow down their rate of development in the wet season and the adult butterflies do not emerge until the rain begins to decrease in intensity. The word aestivation is inappropriate for such an adaptation and I have therefore coined the word 'pluviation' to describe reduced activity in the wet season (Owen, 1971).

Lunar rhythms

A number of marine invertebrates are known to have lunar rhythms, and it has recently been shown that several species of tropical African aquatic insects also show a lunar periodicity in emergence and activity.

In Lake Victoria adults of the mayfly, *Povilla adusta*, emerge in large numbers at the period of the full moon (Hartland-Rowe, 1955). Figure 5.7 shows the days before and after full moon during which 22 different swarms emerged; a marked peak in emergence occurs on the second day after the full moon. Swarms of this species may be recorded simultaneously in areas as much as 80 km apart. The adults live only a few hours and presumably the ecological significance of such synchronized emergence is to ensure that mating is possible.

A number of other aquatic insects show a lunar periodicity in Lake Victoria, including several species of Trichoptera and Diptera (Corbet, 1958). Two species of 'lake fly', *Chaoborus*, emerge and swarm once a month at the time of the new moon. At such times great columns of flies, looking like black smoke, can be seen rising from the lake. The life cycle of these flies takes 2 months to complete, and hence there are two populations of each species separated in time by 1 month (MacDonald, 1956).

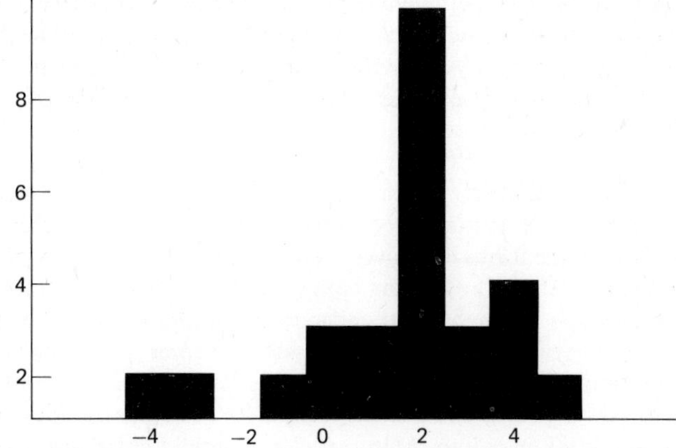

Fig. 5.7 Lunar periodicity in a mayfly, *Povilla adusta*, on Lake Victoria. The number of swarms is shown on the vertical axis, and days before and after full moon (0) on the horizontal axis. (From Hartland-Rowe, 1955.)

Seasonal forms of insects

Several morphological characteristics of tsetse flies, *Glossina* spp., are under climatic control and as a result are more likely to appear at certain times of the year than at others (Glasgow, 1963). These characteristics include size, degree of pigmentation of the abdomen, and certain details of wing venation. Small tsetse appear when food is scarce and also when pupae are exposed to high temperatures. High humidity produces darker flies and high temperature paler flies. Abnormalities in wing venation occur if the pupae are subjected to high temperatures. Glasgow feels that these environmental variations have no survival value. This seems unlikely but it is true that the survival value of such changes is not understood. From a practical point of view these environmentally determined characteristics can be used to make deductions about the sort of climate to which the flies were exposed during their earlier stages.

Several species of African butterflies have wet- and dry-season colour forms that are environmentally determined. In *Precis octavia* (Fig. 5.8) the wet-season form is mainly bright

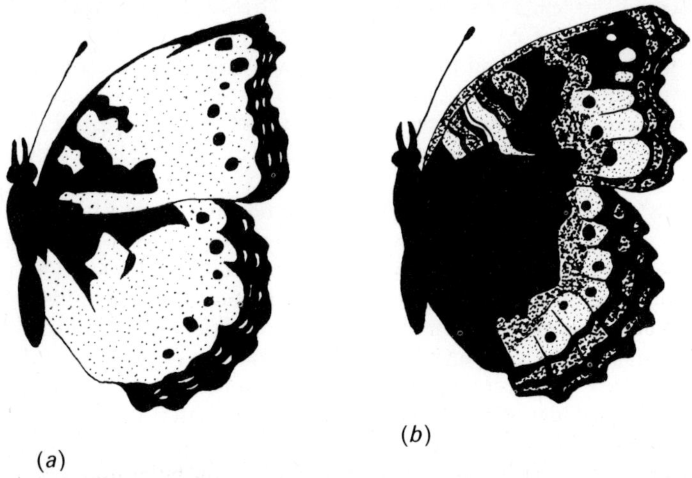

Fig. 5.8 Wet and dry season forms of *Precis octavia*: *(a)* Wet season form; *(b)* Dry season form.

orange with black markings, quite different from the dry-season form which is mainly black with a few blue and red markings. These forms are so different from each other that they could be mistaken for distinct species. In areas where wet and dry seasons alternate and where they are clearly defined, the forms replace each other seasonally. Around Kampala the dry-season form is rare but in parts of West Africa where the dry season is more prolonged it is common. An individual butterfly cannot change colour, but a dry-season female can give rise to wet-season offspring, and *vice versa*. Laboratory experiments have shown that the production of wet- and dry-season forms in butterflies is stimulated by variations in humidity and temperature, but the ecological significance of the forms is elusive.

In the savanna areas of Africa the grass and other vegetation is burnt off during the dry season. This can happen naturally through dry thunderstorms during a drought and may have been occurring for millions of years. Nowadays, and possibly for some thousands of years, most grass fires are started by people, for it is believed (with some justification) that burning the dry grass before the onset of rain improves and hastens the growth of new grass. An area of newly burnt savanna is essentially black in colour. Many species of acridid grasshoppers and other insects have black or blackish forms present among the range of colour types normally found in unburnt habitats, and these forms are much more common on recently burnt ground. Some species are able to change colour from green to black, while in others there are black forms that may seek out an appropriate background, in this case freshly burnt grass. The occurrence of black grasshoppers on burnt ground is seasonal and is associated with burning; and in those species that have been investigated the proximate factors are bright incident light and non-reflecting background. The ultimate factor is probably the deception of potential predators by camouflage.

Daily rhythms

During the course of each 24-hour period an animal undergoes a certain rhythm in its behaviour. Such a rhythm may in some species be endogenous, but in most species external events, particularly changes in temperature, light, and the availability of food, are important. Often a population is highly synchronized in its daily behaviour: one effect of this, especially in sedentary species and in weak-flying insects, is that the chances of mating are greatly increased.

In the savanna elephants tend to move into the shade of trees and bushes between 10 a.m. and 5 p.m., so avoiding direct exposure to the tropical sun. Many diurnal insects do not fly until the dew has dried off the vegetation and the temperature has risen above a certain critical level. Many animals are nocturnal and most species are active for only part of the night. Nocturnal insects are in general most active during the 2 or 3 hours after sunset and to a lesser extent just before sunrise; but there are species that have peaks of activity for other limited periods during the night.

The biting cycles of Diptera have received much attention in tropical Africa because of the medical importance of some species. The East African Virus Research Institute at Entebbe has for some time been investigating activity patterns in biting and other insects. As part of the investigations a high steel tower has been erected in a tropical forest, and insect activity at different levels has been measured quantitatively. Some of the results are published in Haddow *et al.* (1961). A photograph of the tower and the surrounding forest is shown in Fig. 2.1. Swarms of mosquitoes and tabanids have been recorded above the forest canopy in the Zika Forest, Uganda (Corbet and Haddow, 1962). Mosquitoes swarm at sunset while tabanids swarm at sunrise; the significance of the swarms is not known but may well be connected with synchronized mating activity. In the Mpanga Forest, Uganda, the biting cycle of the mosquito, *Mansonia fuscopennata*, varies at different levels in the forest. The insect also makes daily vertical movements (Haddow, 1961).

Diurnal and seasonal rhythms in an African bird

An overall picture of rhythmic events is available for very few species of African animals. One species, a small passerine bird, the black-faced dioch, *Quelea quelea*, has been investigated in detail (Ward, 1965) mainly because of its economic importance. This bird is extremely abundant in the drier parts of tropical Africa. It feeds chiefly on seeds of grasses and at times does an immense amount of damage to cereal crops. In the Lake Chad region of Nigeria these birds roost together in thousands. There is a clear diurnal rhythm in their feeding activity: they leave the roost soon after dawn and fly off to the feeding grounds where they feed for 2 or 3 hours; then they fly back and spend the hot part of the day in the roost; in late afternoon when it becomes cooler they feed again for about 2 hours. Breeding takes place in the wet season which is well-defined and predictable in this area, and at this time of the year the food consists of small insects as well as seeds. With the onset of the dry season larger seeds are eaten and it is at this time that crops are damaged. At the beginning of the rains many grass seeds germinate in the ground simultaneously and the food supply of the birds is suddenly diminished. The birds lay down fat and after a time migrate to areas where the rains have been falling for some weeks and where food has become readily available again. Thus in the black-faced dioch there is a series of adjustments to seasonal and diurnal events, in particular to the food supply. No doubt if information were available similar patterns of behaviour could be described for other species of birds.

Chapter 6

Adaptation to the environment

Natural selection and adaptation

Darwin's theory of natural selection and its adaptive and evolutionary consequences depends on a series of propositions, all of which can be observed and tested. The first of these propositions is that organisms possess the capacity for exponential growth in numbers, but despite this most populations do not grow but remain relatively stable. Death rates, especially among young individuals, are exceedingly high. The second is that organisms are individually variable, no two members of a population are exactly alike, and most individual characteristics are inherited. Thirdly, the probability of death varies between individuals and depends on their genetic make-up.

Natural selection, then, is non-random death. An individual's probability of surviving and reproducing is known as its genetic fitness. Each individual because of its genetic make-up tends to have a different chance of finding food, of being eaten by a predator or attacked by a parasite, or of dying from exposure to adverse weather. Those that survive transmit their genes to their offspring, while the genes of those that die before the age of reproduction are not of course transmitted. The process of natural selection thus results in adaptation to the environment.

Darwin appreciated that natural selection usually produces genetic and adaptive stability; that is to say, generation after generation the genetic characteristics of a population remain unchanged. But if there is a change in the environment other individuals may stand a greater chance of survival and hence different genes tend to be transmitted. Given time this can lead to an adaptive change in the genetic structure of a population, a change that is equivalent to evolution.

Thus besides considering the ecology of populations, communities, and ecosystems it is possible to think in terms of the ecology of genes. The kinds and frequencies of genes are adapted to specific environments, and in this context the study of adaptation to the environment involves the recognition of particular genes in populations and investigating the significance of these genes in the lives of the organisms possessing them. Ecologists cannot afford to neglect natural selection and the capacity of organisms to adapt to new circumstances. As already emphasized, much of tropical Africa has been significantly altered by human activities and it must be assumed that this has produced changes in the genetic structure of animal populations. Moreover, as is well-known, many insect pests and disease-causing organisms show a remarkable capacity to adapt whenever attempts are made to control them by the use of pesticides and other chemicals. In a sense the battle against pests and diseases is a battle that cannot be won but can almost certainly be lost, essentially because of the process of natural selection and the resulting adaptive (evolutionary) changes that it produces.

Polymorphism

Most of the genetically determined characteristics of organisms are controlled by the combined effects of many genes: that is, they are polygenically determined. For instance, birds living in the forested regions of Africa have a darker plumage colour than birds of the same species living in the savanna regions. Many butterflies, too, are darker in West than in East Africa. There are, however, practical difficulties (which can be overcome) in studying polygenic inheritance and its ecological significance, and I shall therefore concentrate on studies of major genes that control easily recognizable and distinct characters. Such genes often (but not always) produce what is called polymorphism within a population.

Polymorphism may be defined as the occurrence together of two or more distinct genetic forms of a species in such proportions that the rarest of them cannot be maintained by recurrent mutation alone. Differences due to age and environmentally determined characteristics are excluded from this definition. The best-known example of polymorphism is the existence of two sexes in a majority of species of animals. Sex is genetically determined and the sex ratio of a population is influenced by natural selection. The relative frequency of males and females is maintained at a level that tends to maximize mating prospects; the sex ratio in most animals is 1:1, but there are some striking and incompletely understood exceptions, as in certain populations of the African butterfly, *Acraea encedon* (Owen, 1971).

Intermediates between forms may occur but by definition they must occur only at rather low relative frequency. Polymorphism, then, is a quantitative concept, involving the recognition of discontinuous variation within populations. Polymorphic forms are normally under the control of genes having clear-cut effects on the phenotype. The existence of polymorphism within a population reflects a balance of selective forces, such that under certain environmental conditions one genotype is at an advantage in terms of chances of survival, while under different conditions another genotype is at an advantage. Such a balance occurs frequently and many examples of polymorphism are known among tropical African animals. Genotypes giving polymorphism may be maintained at stable equilibrium within populations:

1. if the fitness (capacity to survive and reproduce) of the heterozygotes is greater than that of the homozygotes;
2. if the fitness of the genes varies with their frequency in the population, that is to say, a given gene is at a selective advantage over another so long as it does not exceed a specified relative frequency in the population.

The sickle-cell trait in man is one of the best-known examples of a polymorphism maintained by the superior fitness of the heterozygotes over the homozygotes. Individuals homozygous for the trait normally die of anaemia as children but heterozygous individuals are more resistant to malaria caused by *Plasmodium falciparum* than are homozygous non-sicklers. Variations in the frequency of the trait in Africa are largely attributable to variations in the intensity of malaria as a killing disease. Malaria is in part dependent upon concentrations of people and it might be expected that the percentage of sicklers would be higher in areas of high than in areas of low population density. This is so in Liberia where the gene reaches its highest frequency in areas of greatest population density (Livingstone, 1962).

Polymorphism in land snails

The effect of population density on gene frequency has been investigated in the land snail, *Limicolaria martensiana*. This snail occurs in isolated populations and around Kampala, Uganda, there are four distinct colour forms of the shell, as shown in Fig. 6.1. The three pallid forms (b–d) appear to have arisen as dilute mutations of the streaked form, since in most of the pallid shells streaking is just discernible as pale lines. Table 6.1 shows the relative frequency of these four forms and the population density, in ten populations around Kampala. The relative frequency of the streaked form is highest where the population density is low; indeed where the snails are at a density of less than 1 per m^2 almost all are streaked. The pallid forms reach high relative frequencies in populations where the population density is greater. In this species there is

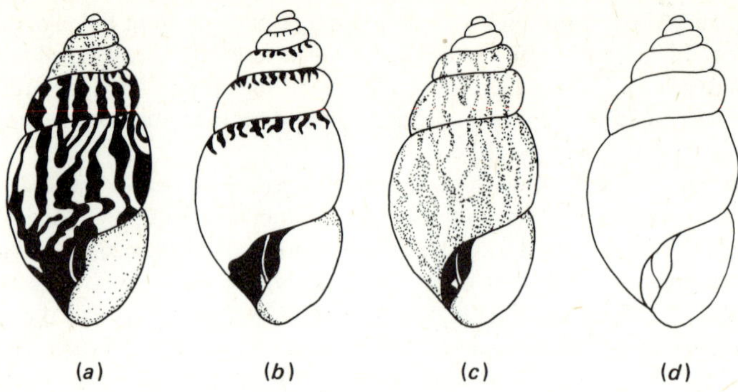

Fig. 6.1 Colour forms of the snail, *Limicolaria martensiana*. The snails are basically pale buff. The black areas shown are dark brown and the shaded areas pale brown. Adult snails are about 3 cm long. *(a)* streaked, *(b)* pallid 1, *(c)* pallid 2, *(d)* pallid 3.

no evidence of heterozygous advantage and I have suggested (Owen, 1963) that the significance of the polymorphism lies in the fact that the colour forms contrast markedly with each other. To the human eye the streaked form appears cryptically coloured; its intricate pattern makes it difficult to see in natural situations, especially where bright sunshine and deep shade alternate. In contrast the pallid forms do not appear cryptic, indeed they often seem conspicuous. There is evidence that a great many predators, especially birds and rodents, eat the snails. Possibly polymorphism in *Limicolaria martensiana* is maintained by what has been termed 'apostatic selection', in which the selective advantage of two or more contrasting colour forms lies in their conspicuous difference from each other. In theory predators, using past experience as a guide in finding their prey, find proportionately more of a colour form with which they have already had significant success. Hence rare colour forms that contrast with the commoner forms would be at a selective advantage because they would tend not to be recognized as prey. In addition, predators would be better able to develop a search-image when the population density of the prey is high than when it is low. In *Limicolaria martensiana* the streaked form may be at an advantage because of its cryptic coloration as long as the population density is not above a certain critical level, at which point contrasting non-cryptic forms assume an advantage simply because they are different.

Table 6.1 *Relative frequency of colour forms of* Limicolaria martensiana *in populations around Kampala, Uganda.*

Population	*Number examined*	*Per cent*				*Density per m²*
		streaked	*pallid 1*	*pallid 2*	*pallid 3*	
1	412	44·4	13·8	36·2	5·6	>100
2	616	55·0	21·0	20·0	3·9	>100
3	1 594	61·4	16·2	19·7	2·7	>100
4	3 455	68·4	15·2	12·9	3·5	26
5	1 049	75·7	10·2	12·3	1·8	21
6	512	77·8	10·7	11·5	—	8
7	411	92·6	7·4	—	—	5
8	130	97·7	2·3	—	—	<1
9	382	100·0	—	—	—	<1
10	75	100·0	—	—	—	<1

Mimicry in butterflies

Another example of the relative fitness of genes depending upon their frequency in the population is found in the polymorphic mimetic butterflies of tropical Africa. But first a consideration of the phenomenon of mimicry. In Africa, especially in the forested regions, a great many species of butterflies are boldly patterned in black and white or black and orange. Many of these butterflies are unpalatable to potential predators and it has long been known that the conspicuous colour patterns function as an advertisement of their distasteful qualities. Certain species of palatable butterflies mimic the colour and pattern of distasteful species to a remarkable degree, and it is assumed that they are often mistaken for unpalatable species by predators and are thus afforded protection. This does not, of course, mean that distasteful or mimetic butterflies are completely immune from the attacks of predators but rather that they enjoy a slight selective advantage over palatable non-mimetic species. Two kinds of mimicry can be distinguished: Batesian mimicry, in which there is a distasteful model and a palatable mimic, and Müllerian mimicry, in which several, often unrelated, distasteful species resemble one another in colour and pattern. In practice it is not always easy to separate the two sorts of mimicry and in the forests of tropical Africa there is an assemblage of species of mimetic butterflies with a wide spectrum of palatability. This assemblage contains many examples of polymorphism.

Polymorphism is usually restricted to the mimics. Thus a species of butterfly may exist in a series of sympatric colour forms, each of which matches closely a different distasteful species of model. Often it is only the females that exhibit mimetic polymorphism, the males being non-mimetic. Among the butterflies of tropical Africa there are two big families of unpalatable species, the Danaidae and the Acraeidae. Not all species in these two families are equally distasteful and even within species there is considerable variation among individuals. A great many other butterflies, particularly in the families Nymphalidae and Papilionidae, are mimics of species in these two families.

The swallowtail, *Papilio dardanus*, is a member of the Papilionidae. The male is yellow with black markings and looks like a normal swallowtail butterfly. In many (but not all) areas the females are strikingly different from the males and are highly polymorphic, each colour form bearing a strong resemblance to a species of Danaidae or Acraeidae. Some of these resemblances are shown in Fig. 6.2. The relative frequency of the colour forms of the mimic is adjusted to the frequency of the available models; mimics are normally rarer than models, but common models have common mimics. If mimics were too common predators would soon learn that they are palatable.

A similar situation prevails in the nymphalid butterfly, *Pseudacraea eurytus*, one of the most polymorphic animals known. It is common in forests throughout tropical Africa. Both males and females are polymorphic and mimetic and each form resembles almost exactly a species in the unpalatable acraeid genus, *Bematistes*. In *Pseudacreaea eurytus*, and probably also in *Papilio dardanus*, the destruction of the forest by human activities seems to have altered the close association between mimics and models. Thus disturbed forest is unfavourable for certain species of *Bematistes* while others are able to exploit it more fully than before. In some areas there is a less precise adjustment between mimics and models than there is presumed to have been in the past (Owen, 1974b).

Polymorphism and background colour

Polymorphic mimicry involves both a qualitative and a quantitative adjustment between two or more species. Similar adjustments occur between polymorphic species and the background colours of the environments in which they live.

In many parts of the world, and especially in the humid tropics, there are two main

Fig. 6.2 Models and polymorphic mimics. Right: three mimetic colour forms of *Papilio dardanus*; left: three species of models (two *Amauris* and one *Bematistes*). The heavily shaded areas are orange and the lightly shaded areas yellowish; all other areas are black or white, as shown.

background colours against which animals may conceal themselves, one green, the other brown. An extraordinary variety of small terrestrial animals are green or brown and many instances of green-brown polymorphism are known, especially in the larvae and pupae of Lepidoptera, in praying mantids, and in the nymphs and adults of grasshoppers. In the long-horn grasshopper, *Homorocoryphus nitidulus*, there are six distinct colour forms: green, brown, green with purple stripes, brown with purple stripes, green with a purple head, and brown with a purple head. Table 6.2 shows the frequency of the six forms in a random sample of about 10 000 specimens collected at Kampala, Uganda. Green is about twice as frequent as brown, and together these two forms account for nearly 97 per cent of the sample. The remaining four forms are relatively

Table 6.2 *Relative frequency of colour forms of* Homorocoryphus nitidulus *at Kampala, Uganda.*

Colour form	*Number examined*	*Per cent*
Green	6 682	63·34
Brown	3 521	33·38
Green with purple stripes	305	2·89
Green with purple head	34	0·32
Brown with purple head	5	0·05
Brown with purple stripes	2	0·02

rare, together comprising 3·28 per cent. The colour forms occur in both sexes but there is a statistical association of green with females and brown with males. *Homorocoryphus nitidulus* lives in grass in somewhat damp and humid places. The green form matches living grass exactly; likewise the brown matches dead grass. The purple on the remaining four forms bears a striking resemblance to the purple areas of anthocyanin pigment that occur on grass leaves and stems in the habitats in which the grasshoppers live. There seems little doubt that in these habitats green is the predominant colour, brown is easily next, and the purple on green and brown leaves and stems in the rarest. Hence the three conspicuous colours in the habitat of the grasshopper are the same colours that occur in the grasshoppers themselves; these colours can be similarly ranked in both the habitat and the grasshopper, suggesting that each of the forms is adapted not only to the colour but also to the amount of that colour in the background.

The ability to change colour may in some animals be superimposed upon polymorphic variation. In the side-striped chameleon, *Chameleo bitaeniatus* (Fig. 6.3), the young are brown,

Fig. 6.3 Side-striped chameleons, *Chameleo bitaeniatus*, lurking among vegetation.

and their ability to undergo reversible colour change is limited to different shades of brown: they cannot change to a very different colour. The adult males are turquoise-blue with a dark reddish-purple dorsal crest, a yellowish gular patch, and a mid-lateral line that lacks pigment, bordered ventrally by a broken reddish-purple line. Adult females lack the conspicuous crest pattern and are basically either green or brown. Reversible colour change occurs in the adults of both sexes: they become darker at lower and paler at higher temperatures, darker in bright daylight and paler at night, and the entire colour and pattern is intensified (especially in the males) upon meeting another chameleon or when attacked by a predator. The males, then, are alike while the females are either green or brown. In some areas green females are about three times as frequent as brown (Ogilvie and Owen, 1964). There is no evidence that adult females can change from green to brown or from brown to green. Hence in this chameleon the females occur in two distinct colour forms and in addition possess an ability to change the intensity of their basic colour with different environmental stimuli. This results in considerable diversity in colour and pattern within a population, especially if, as is often the case, there are differences of light and temperature in different parts of the habitat. Such diversity may afford protection from predators liable to form search images, such as birds.

The effects of isolation

In Chapter 1 I described how geographical isolation could result in the evolution of species. There is no doubt that both geographical and ecological isolation can have fundamental effects upon the likelihood of the occurrence and frequency of particular genes in a population, and therefore on the way the population is adapted to its environment. Even quite small distances can act as barriers to the dispersal of individuals, and therefore of genes, from one population to another. A great many animals, although presumably physically capable of moving from one population to another, in fact do not. Hence some populations geographically close together may be genetically quite different. The snail, *Limicolaria martensiana*, is an example of a species that forms genetically distinct populations; the beetle, *Gnathocera trivittata*, is another, although unlike the snail nothing is known about the significance of the variation between populations.

Gnathocera trivittata occurs in grassy places throughout tropical Africa. The adults feed on the pollen of grasses and during bright sunshine they may be found sitting on the flower heads of grass. These beetles are orange and black, with or without yellow markings on the upper- and underside. The yellow markings vary between individuals. Six forms can be recognized, forming a graded series from 1, the darkest in which the yellow markings are absent, to 6, the palest in which they are conspicuous. The forms are shown in Fig. 6.4. Forms 2 to 5 can be regarded as intermediates between 1 and 6. Table 6.3 shows the relative frequency of the forms in five populations in Uganda. At Kololo and Naguru intermediates are frequent, while at Kaazi, Paraa, and Lake Kitagata intermediates are rare or absent. Kololo and Naguru are adjacent hills separated by 800 m of habitat that is unsuitable for the beetles. Each of these hills has a mode at 1, but the 19·5 per cent difference in the relative frequency of 1 is statistically significant. Hence the two populations are similar in that they both tend towards bimodality in variation but differ in the frequency of the forms. Kaazi is about 16 km from Kololo, and Paraa is about 270 km from Kaazi. Yet Kaazi is very different from Kololo, in that there is a conspicuous single mode at 5 and 6, while at the same time it is very similar to Paraa despite the large distance between these populations. Lake Kitagata, about 280 km west of Kololo, Naguru, and Kaazi, is quite different from any of these populations: there are two modes, one at 1 and the other at 6, but there are no intermediates.

Similar differences between populations of *Gnathocera trivittata* occur in West Africa. The variation is striking and conspicuous (Fig. 6.4), but what is its significance? No one knows, and here lies the challenge.

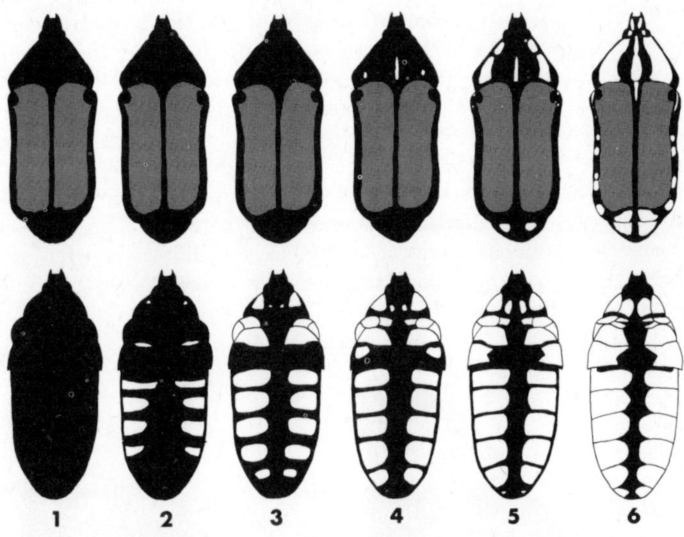

Fig. 6.4 The six colour forms of the beetle, *Gnathocera trivittata*. The shaded areas are yellowish-orange, the white areas pale yellow, and the black areas dark brown.

Ecology, evolution, and taxonomy

It is now realized that many species are much more variable than was previously thought. Thus in the land snail genus, *Limicolaria*, about 170 species have been described for Africa, mostly based upon differences in the colour and pattern of the shell. It is known that the snails in this genus are highly variable and polymorphic and it is likely that once populations have been thoroughly investigated it will be found that there are no more than about a dozen species in tropical Africa. Variation in *Limicolaria* results from adaptation to local environments and from

Table 6.3 *Relative frequency (per cent) of colour forms in populations of the beetle,* Gnathocera trivittata, *in Uganda.*

Population	*Colour form*						*Number examined*
	1	2	3	4	5	6	
Kololo	72·0	4·3	5·5	0·6	11·2	6·4	328
Naguru	52·5	2·0	2·0	5·1	12·1	26·3	99
Kaazi	—	—	1·0	—	16·5	82·5	204
Paraa	—	—	—	—	2·9	97·1	35
Lake Kitagata	62·5	—	—	—	—	37·5	251

the effects of isolation between colonies. The butterfly, *Acraea encedon*, is another colony-forming species that occurs throughout Africa south of the Sahara. Each population can be characterized by the occurrence and frequency of polymorphic forms, by the presence or absence of any one of 20 small black spots on the underside of the hindwing, and by the sex ratio. No two populations are the same and at least one, at Gegbwema in Sierra Leone, appears to be sufficiently isolated and distinct to be in the early stages of speciation: individuals from this population will not breed with those from adjacent populations and yet most *Acraea encedon* from Sierra Leone can be successfully bred (in the laboratory) with individuals from Uganda.

Taxonomists have frequently been confused by the variation found within populations. The butterflies shown in Fig. 6.5 illustrate this particular point. They are females of *Papilio phorcas*, a common forest species. These strikingly different colour forms are found in the same areas. They are controlled by a single gene difference, sex-limited in effect, but the significance of the difference is not known; in this species there is no question of mimetic resemblance. Butterfly collectors could be forgiven for considering the two female forms of *Papilio phorcas* as different species.

Differences between individuals are often neglected by population ecologists but the fact that such variation exists may be of fundamental importance in determining population size and population density. The examples discussed in this chapter will, I hope, draw attention to the relationship between ecology, evolution, and taxonomy, viewed especially in terms of the adaptation of animals to their environment.

Fig. 6.5 The two female forms of *Papilio phorcas*. The shaded areas are bright emerald-green, the white areas pale yellow, and the black areas black. Males are all like the green and black female.

Chapter 7

An introduction to the ecology of man

Whatever special qualities man may attribute to man, there is no doubt that from the standpoint of ecology man is an animal with animal needs. Man is also an abundant animal. And unlike most species, man is increasing in numbers very rapidly. This increase is relatively recent and in tropical Africa has occurred especially in the past 50 to 100 years. The present situation is that some (perhaps most) African countries can expect to double their population in the next 20 or 30 years.

Since 1966, when the first edition of this book was published, ecology has become a household word. In a matter of 10 years people in all walks of life have become aware of the dangers of over-population, famine, malnutrition, and the rapid consumption of non-renewable natural resources. People have starved in the dry areas of the Sahel and in Ethiopia; the predicted economic development of most tropical African countries has been frustrated by soaring prices of raw materials, particularly of oil, and, of course, by the rapid increase in the population. In addition, despite political independence, there has been little economic independence: each African country still relies on the export of raw materials which are consumed by the developed industrial countries. In ecological terms the future of Africa is at best shaky and there are few signs of a radical change for the better. In this chapter I offer a summary of what I believe to be the important features of man's ecology in tropical Africa. I hardly need stress that what I have to say reflects, even more than in the preceding chapters, a personal viewpoint.

Population trends

In mid-1974 the population of the continent of Africa was estimated at just over 411 million. Birth rates were estimated at 47 per 1 000 and death rates at 21 per 1 000; both of these figures are substantially higher than for other continents. In mid-1974 there were about 96 million people in North Africa (Africa north of the Sahara, but including the Sudan) and about 27 million in southern Africa (chiefly in the Republic of South Africa, but including also Botswana, Lesotho, Namibia, and Swaziland). This leaves about 288 million in tropical Africa. The highest populations occur in Nigeria (80 million), Ethiopia (28 million), and Zaïre (24 million).

All population figures are estimates and some are better than others. In industrial countries most births and deaths are officially registered; in Africa the United Nations estimates that only 4 per cent are officially recorded. Counting people is no easier than counting snails; both operations are subject to large and variable margins of error.

Estimates of birth and death rates per thousand enable the calculation of the rate of population growth. In most of tropical Africa this appears to be between 2 and 3 per cent a year, but, because of the lack of reliable censuses it must be assumed that the rate is almost certainly higher than official (national and United Nations) figures. (In most industrial countries the growth rate is well below 1 per cent a year.)

Take, for example, the widely differing census results for Nigeria, the country with the biggest population. According to the latest figures Nigeria's population is now about 80 million – 25 million (or 45 per cent) higher than the 1963 census. Indeed the 1963 census indicated a growth rate of about 2 per cent annually, which is not especially high by tropical African standards. The recent figure of 80 million gives an annual growth rate (over 10 years) of 5 per cent. This is higher than any other country in the world and, if true, probably represents the maximum possible for a human population. It is likely, however, that the 1963 census was a serious under-estimate.

The 1969 censuses in Kenya and Uganda both gave estimates well above what had been expected; indeed almost everywhere there are probably many more people than the official figures suggest. The trouble is that few countries have been able to conduct a reliable census, and for most of them there is simply no estimate worthy of serious consideration.

Why count people? It is known that all populations are eventually limited in size by the carrying capacity of the environment, and also that a growing population can at least occasionally reduce the carrying capacity. This may have happened, for instance, during the recent drought in the Sahel, where both people and cattle suffered not just from the lack of rain and its effect on the vegetation but also from the results of overgrazing and the temporary conversion of savanna into desert.

In a sense, then, each nation offers a carrying capacity to its population, a carrying capacity to be defined essentially in terms of the food it can produce or import. Once the population begins to exceed the capacity, widespread famine can be expected. Most of us would agree that this is something to be avoided and hence the need to obtain vital statistics about population trends.

In Africa the majority of people live in what might be described as rural surroundings. But in recent years there has been a spectacular move to urban areas, triggered, it is believed, by dissatisfaction and the feeling that towns offer more opportunity. In 1957 the population of Kinshasa, Zaïre, was estimated at just over 380 000; 10 years later it was over 900 000, and by now (1975) must be over a million. This kind of increase, which has occurred in virtually all African cities, may be attributed largely to migration from rural areas. It is beyond the scope of this book to discuss the social consequences of such a massive shift in the population; a detailed account is given by Little (1974), and, in more ecological terms, by Owen (1973).

Population and resources

Most species of animals are comparatively restricted in diet and the kind of food eaten by a species throughout its geographical range does not vary enormously. The diet of man, however, varies conspicuously in different areas. In some places people are almost completely herbivorous, while in others they are mainly carnivorous. In many parts of Africa human populations are almost completely dependent upon one or two crops: banana and cassava are two such crops. In other parts of Africa man is a carnivore: some groups of people subsist entirely on domestic animals, especially cattle.

Throughout much of tropical Africa people are dependent upon crops that they grow for themselves. These crops, which include banana, sweet potato, cassava, maize, rice, and a wide variety of fruits, were almost all introduced from other parts of the world, chiefly from tropical America and tropical Asia. It is remarkable that the vast equatorial forest of Africa has not produced a single important plant that can be cultivated for human food. There are of course many edible fruits in the forest, but these are only suitable for gathering and have thus far not been cultivated on a significant scale. The savanna, too, has produced very few crops but an exception must be made of sorghum.

There is usually no overall food shortage in most parts of tropical Africa, although in areas subject to seasonal drought there may be severe shortages extending over several consecutive years. Many of the staple crops grow easily with a minimum of attention, provided traditional methods of diversifying the crops are maintained. Monocultures, usually of cash rather than of staple crops, present more of a problem and are periodically devastated by pests and diseases. People growing food for themselves have accepted and incorporated species of plants introduced from elsewhere but except where there is a substantial profit motive, still maintain the ancient method of diversifying what they grow. This makes good ecological sense if it is acknowledged that in all ecosystems diversity generates stability: outbreaks of pests and diseases are far less likely in mixed stands of crops than in monocultures.

But although severe food shortage may be rather unusual in tropical Africa there are some dangers especially if, as seems likely, the present trends in agriculture are extended. One of these is a shortage of protein, which leads to malnutrition. Another arises from the switch from staple to cash crops for export to industrial countries. Both are further aggravated by the rapid growth of the human population.

Protein malnutrition is largely (but not entirely) due to what is loosely called 'tradition'. That is to say, people tend to eat crops that are deficient in protein even though it would be perfectly possible to grow other crops rich in protein. Thus protein deficiency is decided more by the population's tradition of food preference than by the agricultural potentialities of the area; some of the poorest agricultural areas have little or no protein deficiency. Plantains and cassava, which are staples in many areas, consist almost entirely of carbohydrates with only a fraction of protein, yet the people exist on little else. In southern Uganda and elsewhere, up to 90 per cent of the children show signs of protein deficiency at some stage. The effects of protein deficiency are especially obvious in children who have just been weaned from the breast and have been changed to a diet of carbohydrate: children 1 to 3 years old are highly susceptible. In extreme cases the skin and hair are deficient in melanin pigment and there may be retardation in mental development, which of course is likely to have serious consequences later on. Few children actually die of protein deficiency, but many suffer from it (Fig. 7.1).

For example, in southern Uganda the main meal of the day consists of a large quantity of steamed cooking banana or other carbohydrate staple (Fig. 7.2). Sauces are made from groundnuts, mushrooms, certain kinds of green leaves, tomatoes, and beans, and are served with the steamed banana. Then there are side dishes of eggs, meat, fish, and (when in season) termites and grasshoppers. A meal such as this is extremely filling, but often only very small quantities of sauces and side dishes are available and hence the average meal is deficient in protein. The protein intake could easily be increased by eating more beans, which grow well in southern Uganda, but steamed bananas are much loved and are readily available, and traditions die hard.

In The Gambia (unlike Uganda) there is considerable seasonal change in the availability of the staple crops. Cereals such as findi, maize, rice, and sorghum tend to succeed each other through the year. The diet is predominantly carbohydrate and is supplemented by small amounts of meat, fish, groundnuts, beans, and various seasonal leaves and fruits. At certain times of the year, especially in the dry season, the cultivators have little or no work, while at other times they expend a great deal of energy in preparing the ground for cultivation. When the heavy work starts energy expenditure exceeds intake and body tissue is drawn upon. There is thus a seasonal change in body weight associated with fluctuations in the food supply and the amount of energy expended in preparing the ground for cultivation.

The economy of each tropical African country is uneasily dependent on the export of raw materials to industrial countries. These raw materials are either renewable plant products – e.g., cocoa from Ghana, rubber from Liberia, and groundnuts from Senegal – or non-renewable minerals – e.g., diamonds from Sierra Leone (Fig. 7.3) and copper from Zambia. Nigeria has recently emerged as a major oil producer and it will be interesting to see what effect this has on the nation's economy.

Like the staple crops, most of the cash crops are introductions from elsewhere. More and more land is being converted from forest and savanna and from small family-owned farms into extensive monocultures of cash crops. I have already mentioned that monocultures are highly susceptible to pests and diseases, but even more significant is the uncertainty about the future markets for these products. Some of the more important – cocoa, coffee, tea, and sugar – are to a large extent luxury items and are not essential as human food. If, through economic inflation,

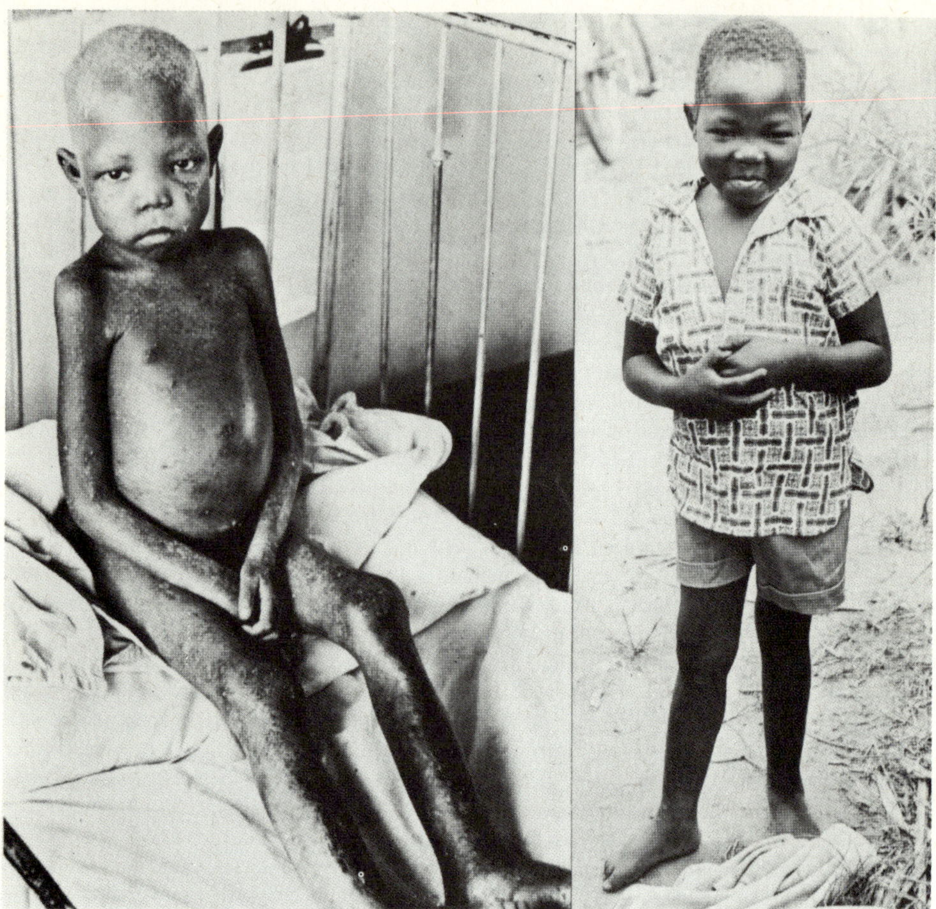

Fig. 7.1 At the time these pictures were taken Oketch was eight years old. He was admitted to hospital suffering from severe protein deficiency. His tribe are traditionally millet-eaters, but at his home they have changed to the more readily available bananas and cassava. As can be seen in the left-hand picture, his face was swollen, his skin was in bad condition, and his limb muscles were under-developed; at this time he was incapable of normal activities. In hospital he was treated with milk for three weeks, and when strong enough to walk was sent home. A week later he was found lying on the floor at his home; all the dried milk given him was stacked in a corner of the house and was unused. He was taken back to hospital and kept for eight weeks. After that, as can be seen in the right-hand picture, he recovered and started school. Malnutrition can often be cured by an adequate supply of milk, as in this case, but could easily be prevented by eating beans, meat, and fish instead of nothing but bananas and cassava. (Photographs by Erasmus Harland.)

people in rich countries were forced to give up certain luxury items, I would predict that cocoa-products would be among the first. And yet a country like Ghana relies heavily on the export of raw cocoa, so heavily, in fact, that it now has to import food to feed the people involved in producing cocoa. The cocoa crop has for years been subjected to ups and downs on the world market; this is what economists sometimes call 'nervousness' in world trade in a particular commodity. Continued dependence on a raw material like cocoa, a product that no one really needs, could easily lead to a highly unstable economy and, it might be added, to a hungry population.

Fig. 7.2 A market scene in tropical Africa. Green cooking bananas, sweet potatoes, and cassava are stacked everywhere. There is cheap food for all, but the food contains little protein. A diet of banana and cassava is the cause of the condition of the little boy shown in Fig. 7.1. (Photograph by Ministry of Information, Uganda.)

What, then, should be done? An ecological approach strongly suggests that diversification together with a maximum of self-sufficiency in essential food production would go a long way towards creating a more stable state of affairs. It remains to be seen, though, if the nations of tropical Africa will reduce their reliance on the export of crop products; thus far it would appear that Tanzania is one of the few nations to adopt officially a policy of self-sufficiency in food production, and even here the future is uncertain because of unreliable rainfall over much of the country and a rapidly expanding population.

Indeed, outsiders view tropical Africa's rapidly expanding population as the main obstacle to raising the standard of living for the majority of the people. But 'standard of living' is a misleading term; more accurately it means 'level of consumption' which may or may not be synonymous with the quality of life, at best an exceedingly difficult parameter to define and measure. Nevertheless there is throughout the world a correlation between level of consumption and the rate of population growth. Low family size is correlated with affluence and this has led to the belief that one way to reduce the rate of population growth is to raise the standard of living. But whether a rise in the standard of living in Africa would lead to a drop in the rate of population increase is a matter of conjecture.

This particular point is further elaborated in Table 7.1 where estimates of birth rates are shown side by side with per capita energy consumption. In each of the seven African countries the annual birth rates are between 40 and 50 per thousand of the population. These rates, which are three times as high as in the United States and the United Kingdom, are among the highest in the world and are typical of all tropical African countries. But the per capita energy consumption in these (and other) African countries is exceedingly low, as can be seen in Table 7.1. For

Fig. 7.3 Diamonds are found in about ten tropical African countries. They are mined and exported and for some countries are an important source of revenue. *(a)* Labour-intensive diamond panning in Sierra Leone, a country that depends heavily on the export of diamonds to industrial countries. How long there will be diamonds to export is a matter of opinion: Sierra Leone may have no more than another five or ten years before it has to find an alternative. *(b)* Mechanical removal of the 'over-burden' in order to expose diamond-bearing ore in South West Africa. This method is extremely destructive to the environment and once the diamonds are gone there will be nothing left but a ruined landscape. (Photographs supplied by De Beers Consolidated Mines Limited.)

example, the world average is about 200 times that of Burundi and more than six times that of the Ivory Coast, a country with one of the highest rates of per capita energy consumption in Africa.

The difference in energy consumption between industrial and tropical African countries arises not so much from differences in food consumption but from the utilization of fossil fuels (oil, coal, and natural gas). The per capita figures shown in Table 7.1 include, of course, energy used

Table 7.1 *Estimated population, birth rate, and per capita energy consumption in selected tropical African countries, and a comparison with industrial countries and the world as a whole.*

	Population size (millions) mid-1974	*Birth rate per thousand of the population*	*Per capita energy consumption in 1972 (in kg of coal equivalent)*
Tropical Africa			
Burundi	3·6	48	10
Chad	4·0	48	20
Ethiopia	27·9	46	35
Ivory Coast	4·7	46	309
Kenya	13·5	49	165
The Gambia	0·4	42	82
Zaïre	24·1	45	86
Industrial countries			
United States	217·5	15	11 611
United Kingdom	56·0	14	5 398
USSR	252·0	18	4 767
World	4 061·1	35	1 984

in such enterprises as 'defence', and in transport and heavy industry. All of these require metals as well as energy. Evidently there is an extraordinary inequality in the utilization of the world's resources which means, at the moment, that an average child born in an industrial country can expect to consume in its lifetime many times the quantity of the world's resources consumed by a child born in, say, Burundi. Thus although birth rates in tropical Africa are high the ecological impact of each birth is relatively small, even by world standards.

A diversity of diseases

It is not my intention in this book to review the extensive literature on the ecology of human disease in tropical Africa; I shall instead draw attention to some of the more important diseases and relate them to the broad ecological scene. There is no doubt that tropical Africa is one of the most disease-ridden areas of the world and that a great many people die every year from diseases that are rare or absent elsewhere. In addition to the diseases that are especially associated with Africa there are others that have been imported from outside. The average person in an industrial country does not suffer from a chronic infection; in Africa a majority of people have at least one infection that persists for a long time, often throughout their lives.

It is possible that until recently human numbers in tropical Africa were limited by disease. In general disease is transmitted more rapidly and its effects are more conspicuous at high than at low population densities; as a result any tendency on the part of a population to increase rapidly is likely to be checked by disease. Many of the diseases that once caused enormous mortality in human populations can now be controlled by the use of drugs and vaccines. Malaria is one such disease; indeed it is likely that malaria was until recently directly or indirectly responsible for the death of a very high percentage of the people in tropical Africa. Much of the mortality from malaria occurs early in life, usually in children less than 5 years old, and often in association with other diseases and with malnutrition. One reason for the rapid rise in human numbers in recent years is because malaria no longer kills so many people in childhood.

But although drugs and vaccines can have conspicuous success with certain diseases, there are others where even now there is insufficient knowledge to control the disease successfully.

Trypanosomiasis (sleeping sickness) is one such disease.* It is transmitted to man and to other mammals (including cattle) by tsetse flies and is perhaps largely responsible for keeping cattle out of the forest region. With the possible exception of locusts and mosquitoes, more is known about the ecology of tsetse flies than other groups of African insects (Fig. 7.4), yet the disease persists

Fig. 7.4 Collecting tsetse flies, *Glossina* spp., from a calf for laboratory research. Much money and effort have been spent on research and resultant attempts to destroy tsetse flies and trypanosomes, but the disease persists in many parts of tropical Africa. (Photograph by British Information Service, Uganda.)

and causes much human suffering. It also prevents the establishment of viable cattle farming over vast areas. Attempts are made to eradicate and control the disease by destroying the tsetse flies and their habitat.

Schistosomiasis (sometimes called bilharzia) is another widespread and common disease in tropical Africa. It is caused by two species of schistosomes, *Schistosoma mansoni* and *S. haematobium*; the intermediate hosts are aquatic snails. *Schistosoma mansoni* causes intestinal bilharzia and occurs in the wetter parts of tropical Africa; in some areas everyone is infected. It does not often kill but accounts for a great deal of lethargy. *Schistosoma haematobium* causes urinary bilharzia and occurs throughout Africa from Egypt to Madagascar and is especially

*Since trypanosomiasis is caused by several different species of trypanosomes it is really a collection of diseases. Thus to speak of it as one disease is a simplification and, as with all simplifications, there is a danger of distorting or even falsifying the true state of affairs. Indeed most diseases, including malaria and influenza, are in reality groups of diseases, a situation that tends to thwart a clear understanding of their ecology and to frustrate attempts to control them.

associated with drier areas. It is uncommon in the forest region but in parts of Egypt it is said that red urine is regarded as normal, so prevalent is the disease. The disease can be cured with drugs, if these are available, but one difficulty is that many people that are infected do not show obvious symptoms. Attempts are being made to destroy the snails that harbour the schistosomes.

Onchocerciasis (river blindness) is caused by infection with a nematode, *Onchocerca volvulus*, which is transmitted by tiny flies of the genus *Simulium*. The disease causes eye defects, including blindness, and skin lesions which may in turn be infected with other diseases. The larvae of *Simulium* are aquatic and occur chiefly in streams. The late-larval stages and the pupae are associated with freshwater crabs and the nymphs of mayflies. Once again attempts to control the disease are directed at the vector and the habitats of the vector.

Malaria, sleeping sickness, bilharzia, and river blindness are four of the more serious chronic diseases that afflict people in tropical Africa. There are many others, and a substantial number of people suffer from one or more of them most of the time. These diseases are persistent and, within an area, do not vary much in intensity, except of course when the environment is disrupted by human activities, as for example with the establishment of an irrigation scheme, an event that has repeatedly resulted in an increase in the incidence of both malaria and bilharzia.

There are, however, other diseases that are not present all the time but which appear and disappear. One of the most dreaded of these is cholera, which recently broke out in West Africa for the first time, and then apparently disappeared just as suddenly. Another is measles, a disease that is nowadays relatively unimportant in Europe, but still a significant, if sporadic, killer in Africa.

In areas where wet and dry seasons alternate the wet season is the worst time of the year for most diseases, but measles is an exception for it nearly always strikes in the dry season. McGregor (1964) gives a graphic account of an epidemic of measles in two Gambian villages. The disease is caused by an air-borne virus and many people associate it with the north wind. An 8-year old child introduced measles into Keneba, one of the villages, after returning from an area where there had been a recent outbreak. The child died soon afterwards and in the meantime the disease had been spread to Jali, a neighbouring village where the first death occurred a little later. Between February and April, at the height of the dry season, 62 of the 437 children in Keneba and Jali died of measles, the death rate of children up to the age of 4 being more than four times that of the older children. With one exception the adults were not affected and it appears that there had been no measles in the area during the previous 12 years. The children could not therefore have acquired natural immunity to the disease.

New epidemic diseases are still being discovered in Africa. Thus in 1959 a virus disease that later became known as 'o'nyong nyong' spread across Uganda into Kenya and Tanzania affecting about 5 million people. The word 'o'nyong nyong' is Acholi for 'joint breaker' which describes the severe joint pains that the disease causes. The disease is transmitted by *Anopheles* mosquitoes of the same species that transmit malaria. It is rarely a killer but when introduced into populations that are already suffering from malnutrition, malaria, bilharzia, and other diseases, it must produce a considerable increase in morbidity.

Many more examples of both chronic and epidemic diseases could be given; there are also inherited diseases, but perhaps because of the prevalence of infectious diseases, these are less understood in Africa than in most parts of the world. An exception must be made of sickle cell anaemia (see Chapter 6) the frequency of which is closely associated with malaria. Much of the genetic variation in human populations, including the abnormal haemoglobins, the blood groups, and certain enzyme polymorphisms, appears to be a selective response to disease pressures, and, in some instances, to climate and diet. The overall picture is of populations constantly exposed to disease which, through the action of natural selection, has resulted in a high degree of immunity. This, in effect, means that people suffer but do not necessarily die. Is this a satisfactory state of affairs? I do not think so, but prospects for a change are not at present good.

The impact of man on natural ecosystems

Thus far in this chapter I have discussed the impact of the African environment on man. In this last section the point can be put the other way round: the effect of man on natural ecosystems.

The most obvious effect of man on other animals and plants is that he destroys and alters ecosystems to suit his own expanding needs (Fig. 7.5). Much of the equatorial forest has now been felled either for timber (much of it exported) or in order to grow crops. Some forest has

Fig. 7.5 The impact of man on the hills of Kigezi, western Uganda. The pale flowers are pyrethrum, which is grown commercially for the manufacture of insecticides. On the right is a plantation of introduced Australian gums, *Eucalyptus* sp. Terraced cultivation may be seen on the hillside in the middle distance. These hills are wet and cool and support a flourishing agriculture and a high human population density. At one time the mountain gorilla occurred on these hills, and is still found in similar but uninhabited areas. (Photograph by Ministry of Information, Uganda.)

since regenerated, to be cut down again, and the area of untouched forest is now quite small. It is only high on mountains or on very steep slopes that the forest is relatively safe from human exploitation. The savanna is grazed by cattle and, if there is enough rain, it is cultivated. Domestic animals have replaced the indigenous grazing mammals in all but a few areas. Swamps have been drained, dams built, and irrigation projects proliferate. There has been much erosion as the result of over-grazing, tree-felling, and cultivation, and much loss of soil fertility. Rich natural plant diversity is increasingly replaced by monotonous monocultures of cash crops, or by mixed assemblages of crops grown by people to feed themselves and their families. In some places quick-growing and productive timber trees have been introduced, including Australian gums, *Eucalyptus*, and even conifers. There are also accidentally introduced weeds like *Lantana camara* and *Tridax procumbens*, both from tropical America and both spreading at an alarming rate right across Africa. But Africa is evidently more resistant to accidental introductions of plants than places like New Zealand and North America and, moreover, rather few foreign animals have successfully established themselves: virtually all of Africa's many species of insects are native.

All of this produces changes in the abundance and diversity of animals. But, for example, an area of forest cleared of most of its trees and planted with coffee is by no means impoverished as far as animals are concerned. Many insects and birds move into clearings created by this kind of agriculture, and a rich, but different, fauna is established. Evidence is beginning to accumulate which suggests that gardens and cultivated areas within the forest region of West Africa are actually richer in certain groups of animals – butterflies, hawk-moths, and snakes are examples – than the forest itself. Ecologists tend to be preoccupied either with agricultural pests and vectors of disease (which is understandable) or with the fauna of natural ecosystems (which is less understandable). The time has come, I believe, for a more thorough investigation of the derived ecosystems created by man. This after all is what Africa is being converted into and the sooner it is understood the better.

The establishment of national parks in East and Central Africa (and to a lesser extent in West Africa) for the preservation of the large numbers of ungulates and their associated predators is sensible and desirable, but has also brought special problems (Fig. 7.6), some of them unexpected. Under protection large mammals tend to increase in numbers, their movements and seasonal migrations are restricted, and in some places they have seriously over-grazed their own habitat. Elephants have contributed to the destruction and retreat of forests, the hippopotamus

Fig. 7.6 In some countries national parks are an important source of revenue and considerable effort is devoted to maintaining a large variety of species of big mammals. This picture shows a white rhinoceros, *Ceratotherium simum* being captured alive and transported from Madi, Uganda for release in the Murchison Falls National Park. It was felt that white rhinoceroses were in danger of being exterminated by poachers in Madi and they were therefore moved to a national park. It was presumed that the ecological conditions in the Murchison Falls Park (where they do not occur naturally) were suitable for white rhinoceroses, but whether they are able to establish themselves permanently there remains to be seen. (Photograph by John Blower.)

is in some places so abundant as to cause erosion and destruction of grassland, and cropping programmes have been started in an attempt to restore numbers to equilibrium. Large predatory mammals have in most places been nearly or completely exterminated. It is said that the destruction of leopards has resulted in a rise in the baboon population but no one knows if this is really so. Some large mammals quickly take advantage of crops grown by man. In Zaïre the chimpanzee feeds largely on the fruits of cultivated paw-paws (introduced from tropical America), some monkeys are destructive in banana gardens, and baboons exploit millet fields. In Sierra Leone chimpanzees have become scavengers of refuse left by man, and also feed on crops. One effect is that longevity of the chimpanzee is increased; another is that there is more rapid dental decay resulting from them feeding from human refuse. In Chad people scavenge food from lion kills, as indeed they do in Uganda; but in Chad people put water out for the lions during the dry season.

One of the most controversial of man's activities in the drier parts of Africa is the periodic burning of grassland. Throughout the savanna, grass is burnt off whenever the weather is dry. It is assumed that burning results in earlier production of fresh grass which is valuable as food for cattle. There are, however, other effects, particularly involving the nitrogen cycle in the savanna ecosystem. Burning and grazing have undoubtedly increased the area of savanna at the expense of forest, and at the southern edge of the Sahara – an area subject to periodic drought – the desert seems to be expanding. Thanks to the work at Lamto in the Ivory Coast the effect of grass fires on animal communities is well-understood; but what has been discovered at Lamto may not be applicable in much drier areas to the north.

Prospects

The words ecology and economics are both derived from the Greek word *oikos* which means a home, a place to which you return and where you are familiar with the local environment. An economic approach to the environment is essentially that of exploitation and profit and is very often short-term in outlook. The establishment of cash crops like tea, coffee, and cocoa in many parts of Africa may initially have seemed a necessary and desirable way of raising standards of living, but it is becoming increasingly apparent that there are both economic and ecological dangers in such enterprises. The ecological approach to man's relationship to the environment differs from the economic in that it is more long-term in outlook.

Human numbers continue to rise steeply and it is estimated that by the year 2000 there will be 860 million people in Africa, perhaps nearly three-quarters of them in tropical Africa. Human demands and needs will also increase and it must be assumed that the people of Africa will expect a bigger share of the world cake than at present, especially in terms of individual energy consumption.

People require food and if famines are to be avoided it is essential that sustained-yield systems of agriculture are maintained and improved. The present shaky dependence on single crop products as a means of earning foreign currency must give way to a more diversified agriculture that produces a variety of cash and staple crops. The attitudes of many outside 'experts'* must be viewed with suspicion. Because of them there has been plenty of research on cash crops, rather little on staples, and even less on traditional and well-tried methods of peasant agriculture. A recent text-book of tropical agriculture (Sutherland, 1971) proclaims, 'What is wrong with subsistence agriculture is that everything that is produced is used up by the people. The people only grow what they need.'

Temperate systems of cultivation often fail when applied to the tropics; no one nowadays would dispute this, but what is to be developed in their place? Here is an urgent need for research

*Although an 'outsider' I would not claim to be an expert, but it will not astonish me if some of the views expressed in this chapter are regarded with suspicion.

and I believe that ecologists could not only contribute but already know some of the answers, at least intuitively. Is it possible, for instance, that insect diversity is a valuable natural resource? The decrease in species diversity that occurs in crop monocultures and the subsequent increase in population size of a few species, leads to outbreaks of pests and the need to use expensive pesticides. This could probably be avoided if monocultures were abandoned and traditional methods incorporating mixed stands of crops were not only encouraged but their ecology more fully understood. A great deal of ecological research in Africa has been and still is oriented towards populations of specific pests, diseases, weeds, and crops. The time has come, I suggest, to adopt a more integrated approach, an approach that makes full use of the ecosystem concept. There is little wisdom in trying to manipulate part of a system without reference to or knowledge of the system as a whole.

In ecological as well as in economic and political terms Africa is at a turning point in its history. The grand design of forests and savannas is fast disappearing and is being replaced by a complex mosaic of cultivation and remnants of the ancient ecosystems. In some areas there are already signs that this mosaic is unstable and fragile in terms of what it can offer the human population: the recent famines in the Sahel and in Ethiopia are examples of what could become more frequent. Information is urgently needed and it is to be hoped that the many excellent universities and research organizations in tropical Africa will continue to expand studies of all aspects of the relationship between man and the environment. And if, on occasion, the ecologist has to turn politician, so much the better.

Chapter 8

Further study

There was a time when school and university students in Africa knew more about European plants and animals than African ones. Happily this situation is now remedied. There is, however, always a lag between the publication of the results of scientific research and their appearance in a more accessible form in books. For many years research organizations and universities in Africa have been producing excellent scientific work but only in the last 10 to 15 years have the fruits of this research become readily available to all.

Some of the text-books now available to biology students in Africa are, in my view, superior to those used in Europe and America. Biology teaching in Africa is developing a sound ecological basis with emphasis on individual participation through field observation and experimentation. For example, *Ecological Biology* edited by D. W. Ewer and J. B. Hall (Longman, 1972), contains a wealth of tropical material much of it derived from research initiated in Africa. It is intended for senior school students and first-year university students, and in particular emphasizes the value of thinking in ecological terms.

The Zoology of Tropical Africa by J. L. Cloudsley-Thompson (Weidenfeld and Nicolson, 1969) treats the subject in terms of adaptations to habitats and climate, and raises many questions about the physiological adjustments of animals to drought, heat, and humidity. C. I. O. Olaniyan's *An Introduction to West African Animal Ecology* (Heinemann, 1968) deals especially with the ecology of coastal habitats and mangrove swamps, topics that I have excluded from my own book, and contains numerous drawings of the animals likely to be found in these and other habitats. Animal ecologists must also be botanists and certainly need to integrate the study of plant and animal ecology. Learning the names of plants is by no means easy – there are so many species – but there are for most parts of Africa excellent guides to the identification of plants and trees. The ecology of tropical forest is discussed in P. W. Richard's *The Tropical Rain Forest* (Cambridge University Press, 1952), and by K. A. Longman and J. Jeník in *Tropical Forest and its Environment* (Longman, 1974), while in *Forest and Savanna* (Heinemann, 1965) B. Hopkins compares and contrasts Africa's two most important ecosystems. A series of essays (in French and English) published under the title *Conservation of Vegetation in Africa South of the Sahara* (Uppsala, 1968) and edited by I. and O. Hedberg persuasively puts the case for the need to preserve what little is left of Africa's natural vegetation. The essays on plant communities high on tropical mountains and in arid areas are especially interesting and the book contains some fine photographs of plant associations.

The similarities and differences between the flora and fauna of Africa and South America are discussed in *Tropical Forest Ecosystems in Africa and South America: a comparative Review* (Smithsonian Institution Press, 1973). The book is edited by B. J. Meggers, E. S. Ayensu, and W. D. Duckworth and contains articles on mammals, birds, amphibians and reptiles, fish, and on a few groups of insects, as well as more general comparisons. R. H. Carcasson's article 'A preliminary survey of the zoogeography of African butterflies' (*East African Wildlife Journal*, Vol. 2, pp. 122–157, 1964) is a useful attempt at interpreting the present distribution of butterflies in terms of past climatic changes. A similar approach has been tried with other groups, including birds, both for Africa as a whole and for various regions within Africa.

Large mammals have attracted more attention than other groups – except of course pests like locusts and disease-carrying insects like tsetse flies – and there are detailed monographs on the natural history and behaviour of such species as chimpanzee, gorilla, and elephant. There is, however, far less information on the interactions between the myriads of small animals, most of them insects, that occur, for example, in a garden or in a tiny patch of cultivated land. Many of these species are predators and their behaviour, as well as the behaviour of their prey, is described by M. Edmunds in *Defence in Animals* (Longman, 1974), a book that draws heavily on observations made in the man-modified environment surrounding the University of Ghana. There is, in fact, no need to travel to remote areas of undisturbed forest and savanna for ecological investigations. The best places are compounds and gardens where it is possible to make regular observations over long periods.

There are plenty of books and articles about the impact of man on the African environment. A disaster like the Sahel drought inevitably produces a flood of publications explaining what went wrong and what should be done. Agricultural problems, too, are well-covered – there are extensive monographs on all the important tropical crops. Publications on tropical agriculture (almost invariably written by 'outsiders') fall into two broad categories: those that offer criticism and advice; and those that describe and analyse events in more ecological terms. As an example of a book in the first category I suggest R. Dumont's *False Start in Africa* (Deutsch, 1966), and in the second category M. P. Miracle's *Agriculture in the Congo Basin* (University of Wisconsin Press, 1967). Few people that write about tropical agriculture have themselves tried to cultivate crops in the tropics. Yet some have strong views, as for example D. H. Janzen in his article 'Tropical Agroecosystems' published in 1973 the magazine *Science* Vol. 182, pp. 1212–1219. Janzen believes that the poor state of tropical agriculture arises from misunderstanding and mismanagement of the new ecosystems that have been created, and that a new, more socially integrated approach is required. All well and good but with this as with all human problems there is the seemingly insurmountable difficulty of turning ideas into practice.

The most important sources of information on animal ecology in tropical Africa are the annual reports and scientific articles produced by research organizations and universities. There is, of course, much variation in the quality and quantity of research in different countries; much depends on the size of the country and its financial resources, but every country produces something.

Glossary

The terms below are defined in the way they are used in this book; some terms have other meanings in a different context.

aestivation: reduced activity of an animal associated with metabolic adjustments to hot or dry weather.
allele: an alternative form of a gene at a given locus on a chromosome.
anadromesis: up-river migration of fish.
apostatic selection: natural selection for genetic traits that stand out from or contrast with the normal.
Batesian mimicry: the resemblance of a palatable mimic to an unpalatable model.
biomass: the product of the number of individuals in a population and their mean weight.
climax community: the final assemblage of plants and animals that becomes established in an area.
community: the assemblage of plants and animals in an area.
competition: an interaction of individuals or species resulting from utilization of essential resources that are in short supply.
consumer: an organism that feeds on living organisms.
decomposer: an organism that obtains its nutrients from dead organic matter.
delayed implantation: a delay between fertilization and the implantation of the embryo in the uterus of a mammal.
density-dependent events or factors: ecological events or factors whose effect varies with the density of the population.
density-independent events or factors: ecological events or factors whose effect does not vary with the density of the population.
dispersion: the quantitative distribution of organisms in an area.
ecological segregation: the tendency for different species to differ in their ecological requirements.
ecosystem: the interrelated organisms in an environment. The concept of the ecosystem incorporates the sum total of interactions between organisms and their interactions with the non-living part of the environment.
endogenous rhythm: a physiological rhythm determined by the internal condition of the organism.
evolution: a cumulative inheritable change in a population.
exogenous rhythm: a physiological rhythm in an organism determined by external environmental factors.
fitness: the relative ability of an organism to transmit its genes to the next generation.
gene: a hereditary unit which through transcription has a specific effect and which can mutate to alternative forms.
genotype: the genetic constitution of an organism.
heterozygote: an individual with two dissimilar alleles (alternative forms of a gene) at a given locus.
homozygote: an individual with two identical alleles (alternative forms of a gene) at a given locus.
irruption: an irregular mass movement of a population.
lunar periodicity: periodic fluctuations in biological events associated with a particular phase of the moon.
migration: a regular movement by individuals or populations to and from a breeding area.
mimicry: superficial resemblance of organisms of different species such that one or both benefits.
Müllerian mimicry: the resemblance of several, often unrelated, unpalatable species.
mutation: a sudden and haphazard change in a gene.
natural selection: the non-random elimination of individuals (and therefore of genes) from a population; often abbreviated to *selection*.
niche: the particular environment in which an organism lives and the way in which it utilizes this environment.
photosynthesis: a process by which green plants capture radiant energy from the sun and convert it to stored chemical energy.
pluviation: reduced activity of an animal during wet weather.
polygenic: of a mode of inheritance: determined by several or many genes.

polymorphism: the occurrence together in the same population of two or more distinct genetic forms in such proportions that the rarest of them cannot be maintained by recurrent mutation alone.
polyoestrus: of female mammals: cyclical changes in behaviour that permit mating more than once a year.
population: a group of organisms of the same species living together.
producer: an organism, normally a green plant, capable of photosynthesis.
productivity: the rate at which chemical energy is produced during photosynthesis.
proximate factors: environmental factors that stimulate periodic events in the lives of organisms; 'biological triggers'.
search image: a predator's pictorial memory of the appearance of prey.
sex-limited: of a gene present in both sexes but expressing itself in one sex only.
sex ratio: the relative proportion of males and females of a specified age group in a population.
speciation: an evolutionary process leading to the formation of species.
species diversity: the number of species present in an area in relation to their abundance; also simply the number of species present.
statistical significance: of a numerical result that is unlikely to have been arrived at by chance.
succession: sequential changes that occur in a community following disruption of the environment.
taxonomy: the study of naming and classifying organisms.
territory: an area defended by an individual against other members of the same species.
trophic level: feeding level, but including also the acquisition of energy by plants.
trophic model: an attempt to represent the working of an entire ecosystem; most trophic models combine assumption with fact.
ultimate factors: environmental factors that determine why an organism evolves the habit of breeding, migrating, etc., at a certain time.

References

Barbault, R. 1971. Recherches écologiques dans la savane de Lamto (Côte-d'Ivoire): production annuelle des populations naturelles du lézard *Mabuya büttneri* (Matschie). *Terre Vie*, **25**: 203–17.

Beadle, L. C. 1933. Scientific results of the Cambridge expedition to the East African lakes, 1930–1. 13. Adaptation to aerial respiration in *Alma emini* Mich., an oligochaet from East African swamps. *J. Linn. Soc.*, **38**: 347–50.

Beadle, L. C. 1961. Adaptations of some aquatic animals to low oxygen levels and to anaerobic conditions. *Symp. Soc. exp. Biol.*, **15**: 120–31.

Bere, R. M. 1962. *The Wild Mammals of Uganda and Neighbouring Regions of East Africa*. Longman, London.

Berrie, A. D. and Visser, S. A. 1963. Investigations of a growth-inhibiting substance affecting a natural population of freshwater snails. *Physiol. Zoöl.*, **36**: 167–73.

Bourlière, F. 1963. Observations on the ecology of some large African mammals. *African Ecology and Human Evolution*, **36**: 43–54.

Bourlière, F. and Hadley, M. 1970. The ecology of tropical savannas. *Ann. Rev. Ecol. Syst.*, **1**: 125–52.

Brooks, J. L. 1950. Speciation in ancient lakes. *Quart. Rev. Biol.*, **25**: 131–76.

Brown, L. H. 1955. Supplementary notes on the biology of the large birds of prey of Embu District, Kenya Colony. *Ibis*, **97**: 38–64.

Buechner, H. K. 1961. Territorial behavior in Uganda kob. *Science*, **133**: 698–9.

Chapin, J. P. 1932. The birds of the Belgian Congo, Part 1. *Bull. Amer. Mus. nat. Hist.*, **65**: 1–736.

Chapman, B. M. and Chapman, R. F. 1964. Observations on the biology of the lizard *Agama agama* in Ghana. *Proc. zool. Soc. Lond.*, **143**: 121–32.

Corbet, P. S. 1958. Lunar periodicity of aquatic insects in Lake Victoria. *Nature, Lond.*, **182**: 330–1.

Corbet, P. S. and Haddow, A. J. 1962. Diptera swarming high above the forest canopy in Uganda, with special reference to Tabanidae. *Trans. R. ent. Soc. Lond.*, **114**: 267–84.

Cott, H. B. 1961. Scientific results of an inquiry into the ecology and economic status of the Nile crocodile (*Crocodilus niloticus*) in Uganda and Northern Rhodesia. *Trans. zool. Soc. Lond.*, **29**: 211–356.

Estes, R. 1975. *Xenopus* from the Palaeocene of Brazil and its zoogeographic importance. *Nature, Lond.*, **254**: 48–50.

Evans, M. E. G. and Blower, J. G. 1973. A jumping millipede. *Nature, Lond.*, **246**: 427–8.

Fryer, G. 1956. A report on the parasitic Copepoda and Branchiura of the fishes of Lake Nyasa. *Proc. zool. Soc. Lond.*, **127**: 293–344.

Fryer, G. 1959. The trophic interrelationships and ecology of some littoral communities of Lake Nyasa with especial reference to the fishes, and a discussion of the evolution of a group of rock-frequenting Cichlidae. *Proc. zool. Soc. Lond.*, **132**: 153–281.

Fryer, G. 1961. Observations on the biology of the cichlid *Tilapia variabilis* Boulenger in the northern waters of Lake Victoria (East Africa). *Revue Zool. Bot. afr.*, **64**: 1–33.

Gillett, J. D. 1965. Analysis of the overall laying-cycle in a population of insects. *Proc. 12th. Int. Congr. Ent. Lond.*, **1964**: 789–90.

Gillon, D. 1970. Recherches écologiques dans la savane de Lamto (Côte d'Ivoire): les effets du feu sur les arthropodes de la savane. *Terre Vie*, **24**: 80–93.

Gillon, D. and Pernes, J. 1968. Etude de l'effet du feu de brousse sur certains groupes d'arthropodes dans une savane préforestière de Côte-d'Ivoire. *Ann. Univ. Abidjan*, **1(E)**: 113–97.

Glasgow, J. P. 1963. *The Distribution and Abundance of Tsetse*. Pergamon Press, London.

Gotwald, W. H. 1974. Predatory behavior and food preferences of driver ants in selected African habitats. *Ann. ent. Soc. Amer.*, **67**: 877–86.

Green, J. 1960. Zooplankton of the River Sokoto: the freshwater medusa *Limnocnida*. *Proc. zool. Soc. Lond.*, **135**: 613–18.

Green, J. 1964. The numbers and distribution of the African fish eagle *Haliaëtus vocifer* on the eastern shores of Lake Albert. *Ibis*, **106**: 125–8.

Greenwood, P. H. 1958. Reproduction in the East African lung-fish *Protopterus aethiopicus* Heckel. *Proc. zool. Soc. Lond.*, **130**: 547–67.

Haddow, A. J. 1952. Field and laboratory studies on an African monkey, *Cercopithecus ascanius* Matschie. *Proc. zool. Soc. Lond.*, **122**: 297–394.

Haddow, A. J. 1961. Entomological studies from a high tower in Mpanga Forest, Uganda. VII. The biting behaviour of mosquitoes and tabanids. *Trans. R. ent. Soc. Lond.*, **113**: 315–35.

Haddow, A. J., Corbet, P. S., Gillett, J. D., Dirmhirn, I., Jackson, T. H. E., and Brown, K. W. 1961. Entomological studies from a high tower in Mpanga Forest, Uganda. *Trans. R. ent. Soc. Lond.*, **113**: 249–368.

Hartland-Rowe, R. 1955. Lunar rhythm in the emergence of an ephemeropteran. *Nature, Lond.*, **176**: 657.

Howes, N. H. and Wells, G. P. 1934. The water relations of snails and slugs. *J. exp. Biol.*, **11**: 327–51.

Jackson, T. H. E. 1961. Entomological studies from a high tower in Mpanga Forest, Uganda. IX. Observations on Lepidoptera (Rhopalocera). *Trans. R. ent. Soc. Lond.*, **113**: 346–50.

Johnels, A. G. and Svensson, G. S. O. 1954. On the biology of *Protopterus annectens* (Owen). *Ark. Zool.*, **7**: 131–64.

Kahl, M. P. 1971. Food and feeding behavior of openbill storks. *J. Ornith.* **112**: 21–35.

Keay, R. W. J. 1959. *Vegetation Map of Africa.* Oxford University Press, London.

Kellas, L. M. 1955. Observations on the reproductive activities, measurements, and growth rate of the dikdik (*Rhynchotragus kirkii thomasi* Neumann). *Proc. zool. Soc. Lond.*, **124**: 751–84.

Kruuk, H. and Sands, W. A. 1972. The aardwolf (*Proteles cristatus* Sparrman) as predator of termites. *E. Afr. Wildl. J.*, **10**: 211–27.

Kühme, W. 1965. Communal food distribution and the division of labour in African hunting dogs. *Nature, Lond.*, **205**: 443–4.

Lamotte, M., Aguesse, P. and Roy, R. 1962. Données quantitatives sur une biocoenose ouest-africaine: la prairie montagnarde du Nimba (Guinée). *Terre Vie*, **16**: 351–70.

Lavelle, P. 1971. Recherches écologiques dans la savane de Lamto (Côte-d'Ivoire): production annuelle d'un ver de terre *Millsonia anomala* Omodeo. *Terre Vie*, **25**: 240–54.

Leston, D. and Hughes, B. 1968. The snakes of Tafo, a forest cocoa-farm locality in Ghana. *Bull. Inst. fr. Afr. Noire*, **30**: 737–70.

Little, K. 1974. *Urbanization as a social process.* Routledge & Kegan Paul, London.

Livingstone, D. A. 1967. Postglacial vegetation of the Ruwenzori Mountains in equatorial Africa. *Ecol. Mon.*, **37**: 25–52.

Livingstone, F. B. 1962. Population genetics and population ecology. *Amer. Anthrop.*, **64**: 44–53.

MacDonald, W. W. 1956. Observations on the biology of chaoborids and chironomids in Lake Victoria and on the feeding habits of the 'elephant-snout fish' (*Mormyrus kannume* Forsk.). *J. Anim. Ecol.*, **25**: 36–53.

Mackworth-Praed, C. W. and Grant, C. H. B. 1952. *Birds of Eastern and North Eastern Africa*, vol. 1. Longman, London.

Madge, D. S. 1965. Leaf fall and litter disappearance in a tropical forest. *Pedobiol.*, **5**: 273–88.

Marlier, G. 1955. Un trichoptère pélagique nouveau du lac Tanganika. *Revue Zool. Bot. afr.*, **52**: 150–5.

Marshall, A. J. and Williams, M. C. 1959. The pre-nuptial migration of the yellow wagtail (*Motacilla flava*) from latitude 0°04′N. *Proc. zool. Soc. Lond.*, **132**: 313–20.

McGregor, I. A. 1964. Measles and child mortality in The Gambia. *W. Afr. med. J.*, **13**: 251–7.

Mead, A. R. 1961. *The Giant African Snail: A Problem in Economic Malacology.* University of Chicago Press, Chicago.

Menzies, J. I. 1963. The climate of Bo, Sierra Leone, and the breeding behaviour of the toad, *Bufo regularis. J. W. Afr. Sci. Ass.*, **8**: 60–73.

Menzies, J. I. 1966. The snakes of Sierra Leone. *Copeia*, **1966**: 169–79.

Moreau, R. E. 1948. Ecological isolation in a rich tropical avifauna. *J. Anim. Ecol.*, **17**: 113–26.

Moreau, R. E. 1952. Africa since the mesozoic: with particular reference to certain biological problems. *Proc. zool. Soc. Lond.*, **121**: 869–913.

Moreau, R. E. 1954. The distribution of African evergreen-forest birds. *Proc. Linn. Soc. Lond.*, **165**: 35–46.

Moreau, R. E. 1963. The distribution of tropical African birds as an indicator of past climatic changes. *African Ecology and Human Evolution*, **36**: 28–42.

Mutere, F. A. 1965. Delayed implantation in an equatorial fruit bat. *Nature, Lond.*, **207**: 780.

Mutere, F. A. 1967. The breeding biology of equatorial vertebrates: reproduction in the fruit bat, *Eidolon helvum*, at latitude 0°20′N. *J. Zool., Lond.*, **153**: 153–61.

Norris, M. J. 1962. Diapause induced by photoperiod in a tropical locust, *Nomadacris septemfasciata* (Serv.). *Proc. Ass. appl. Biol.*, **50**: 600–603.

North, M. E. W. 1963. Breeding of the black-headed herons at Nairobi, Kenya, 1958–62. *J. East Afr. nat. Hist. Soc.*, **24**: 33–63.

Odhiambo, T. R. 1958. Drosophilidae (Dipt.) breeding in cercopid (Hem.) spittle masses. *Entomologist's mon. Mag.*, **94**: 17.
Ogilvie, P. W. and Owen, D. F. 1964. Colour change and polymorphism in *Chameleo bitaeniatus*. *Nature, Lond.*, **202**: 209–10.
Owen, D. F. 1963. Polymorphism and population density in the African land snail, *Limicolaria martensiana*. *Science*, **140**: 666–7.
Owen, D. F. 1964. Bimodal occurrence of breeding in an equatorial land snail. *Ecology*, **45**: 862.
Owen, D. F. 1969. The migration of the yellow wagtail from the equator. *Ardea*, **57**: 77–85.
Owen, D. F. 1971. *Tropical Butterflies: the Ecology and Behaviour of Butterflies in the Tropics with Special Reference to African Species*. Clarendon Press, Oxford.
Owen, D. F. 1973. *Man's Environmental Predicament: An Introduction to Human Ecology in Tropical Africa*. Oxford University Press, London.
Owen, D. F. 1974a. *What is Ecology?* Oxford University Press, London.
Owen, D. F. 1974b. Exploring mimetic diversity in West African forest butterflies. *Oikos*, **25**: 227–37.
Owen, D. F. and Owen, J. 1974. Species diversity in temperate and tropical Ichneumonidae. *Nature, Lond.*, **249**: 583–4.
Owiny, A. M. 1974. Some aspects of the breeding biology of the equatorial land snail *Limicolaria martensiana* (Achatinidae: Pulmonata). *J. Zool., Lond.*, **172**: 191–206.
Phillips, J. 1959. *Agriculture and Ecology in Africa*. Faber and Faber, London.
Reilly, C. 1967. Accumulation of copper by some Zambian plants. *Nature, Lond.*, **215**: 667–8.
Salt, G. 1954. A contribution to the ecology of Upper Kilimanjaro. *J. Ecol.*, **42**: 375–423.
Schaller, G. B. 1963. *The Mountain Gorilla*. University of Chicago Press, Chicago.
Stewart, D. R. M. and Stewart, J. 1963. The distribution of some large mammals in Kenya. *J. East Afr. nat. Hist. Soc.*, **24**: 1–52.
Sutherland, J. A. 1971. *Introduction to Tropical Agriculture*. Angus and Robertson, London.
Swynnerton, G. H. and Hayman, R. W. 1950–51. A check list of the land mammals of the Tanganyika Territory and the Zanzibar Protectorate. *J. East Afr. nat. Hist. Soc.*, **20**: 274–392.
Thiollay, J.-M. 1970. L'exploitation par les oiseaux de essaimages de fourmis et termites dans une zone de contact savane-forêt en Côte-d'Ivoire. *Alauda*, **38**: 255–73.
Thiollay, J.-M. 1973. Place des oiseaux dans les chaînes trophiques d'une zone préforestière en Côte-d'Ivoire. *Alauda*, **41**: 273–300.
Thurston, J. P. 1968. The frequency distribution of *Oculotrema hippopotami* (Monogenea: Polystomatidae) on *Hippopotamus amphibius*. *J. Zool., Lond.*, **154**: 481–5.
Thurston, J. P. and Laws, R. M. 1965. *Oculotrema hippopotami* (Trematoda: Monogenea) in Uganda. *Nature, Lond.*, **205**: 1127.
Wallace, A. R. 1878. *Tropical Nature and other Essays*. Macmillan, London.
Ward, P. 1965. Feeding ecology of the black-faced dioch *Quelea quelea* in Nigeria. *Ibis*, **107**: 173–214.
Wasawo, D. P. S. 1959. A dry season burrow of *Protopterus aethiopicus* Heckel. *Revue Zool. Bot. afr.*, **60**: 65–71.
Williams, C. B. 1951. The migrations of libytheine butterflies in Africa. *Nig. Field*, **16**: 152–9.
Williams, C. B. 1964. *Patterns in the Balance of Nature and Related Problems in Quantitative Ecology*. Academic Press, London.

Index

(**Bold type** numerals are references to Plates)